The Minerals, Metals & Materials Series

Shadia Ikhmayies · Jian Li ·
Carlos Mauricio Fontes Vieira ·
Jean Igor Margem · Fabio de Oliveira Braga
Editors

Green Materials Engineering

An EPD Symposium in Honor of Sergio Monteiro

TMS

Springer

Editors
Shadia Ikhmayies
Al Isra University
Amman, Jordan

Jian Li
CanmetMATERIALS
Hamilton, ON, Canada

Carlos Mauricio Fontes Vieira
State University of Norte Fluminense
Campos dos Goytacazes, Brazil

Jean Igor Margem (Deceased)
Higher Teaching Institute
Campos dos Goytacazes, Brazil

Fabio de Oliveira Braga
National Service of Industrial
Apprenticeship (SENAI)
Rio de Janeiro, Brazil

ISSN 2367-1181 ISSN 2367-1696 (electronic)
The Minerals, Metals & Materials Series
ISBN 978-3-030-10382-8 ISBN 978-3-030-10383-5 (eBook)
https://doi.org/10.1007/978-3-030-10383-5

Library of Congress Control Number: 2018964935

Cover illustration: Cover image from Chapter 'Natural Fibers Reinforced Polymer Composites Applied in Ballistic Multilayered Armor for Personal Protection—An Overview', Sergio Neves Monteiro et al., pages 33–47, Figure 8: Capture of ceramic/projectile fragments. https://doi.org/10.1007/978-3-030-10383-5_4

This Springer imprint is published by the registered company Springer Nature Switzerland AG
The registered company address is: Gewerbestrasse 11, 6330 Cham, Switzerland

Preface

Green engineering is the design, use of processes, and development of products that conserve natural resources, reduce pollution, and exert the smallest possible impact on the environment. Green engineering often promotes sustainability and minimizes risk to human health without immolating economic feasibility and efficiency. So, green engineering is not actually an engineering discipline in itself, but a comprehensive framework for all engineering disciplines. The Green Materials Engineering symposium is a TMS Extraction and Processing Division (EPD) symposium, sponsored by the Characterization of Minerals, Metals, and Materials and the Biomaterials Committees. This symposium is held in honor of Professor Sergio Monteiro from the Military Institute of Engineering, IME, Brazil. The symposium focuses on green materials including natural composites, bio-inspired armors, waste, clays added ceramics, lignocellulosic fibers, biodegradable polymers, and any type of natural material that could be related to engineering applications.

The Green Materials Engineering symposium held during the TMS 2019 Annual Meeting in San Antonio, Texas, USA, received 54 abstract submissions from different research groups, of which 34 were accepted as oral presentations and 20 accepted as posters. Of the presented papers, 33 are published in this book after being peer reviewed. These papers cover different fields including sustainable clays and ceramics, natural fiber composites, nano- and microgreen composites, properties and characterization of green materials, and biomass in armor composites. These materials are characterized using highly sophisticated techniques to examine their microstructure, mechanical, thermal, and functional properties.

This book will appeal to people from academia and industry who are interested in green engineering, sustainability, recycling, and environment, and it can help academia emphasize pollution prevention and incorporate risk into green engineering courses. It helps environmental and chemical engineers; postdoctoral, graduate and undergraduate students; people from industry; and environmental scientists to convert concepts of green engineering and sustainability to real designs, using the most valuable quantitative design tools and performance metrics.

The editors of this book express their genuine thanks and gratitude to TMS for giving the organizers and committee sponsors the opportunity to publish a stand-alone volume for this symposium. The editors also thank the publisher, Springer, who produced the book, and the authors, who are the basis of this scientific work.

Shadia Ikhmayies
Jian Li
Carlos Mauricio Fontes Vieira
Fabio de Oliveira Braga

Contents

Part IV Properties and Characterization of Green Materials

Part V Biomass in Armor Composites

About the Honoree

Sergio Neves Monteiro, born on June 24, 1943, in Rio de Janeiro, Brazil, obtained his bachelor degree in Metallurgical Engineering at the Federal University of Rio de Janeiro, UFRJ, in 1966. Both his M.Sc. (1968) and Ph.D. (1972) degrees were obtained in the Materials Science and Engineering Department at the University of Florida, USA. He also has taken an extension course in energy at the War College in Rio de Janeiro (1975) and a postdoctoral term at the University of Stuttgart, Germany (1976). He joined the staff of the Department of Metallurgical and Materials Engineering at the UFRJ in 1968 and became Professor in 1975; he then retired 18 years later. In 1993, he accepted an invitation to create the Laboratory of Advanced Materials of the State University of the North Rio de Janeiro (UENF) where he is still an active professor. Presently, he is a collaborator professor at the Military Institute of Engineering (IME).

He is the author of over 1,000 papers in the fields of plastic deformation and temperature behavior, mechanical properties of metals and alloys, continuing engineering education, environmental engineering, advanced ceramics, natural fibers, natural fiber polymeric composites, alternative energies, technological programs, and administration on science and technology. He was responsible for over 50 engineering projects and consulting works for the industry. He has lectured over 160 undergraduate, postgraduate, and extension courses, both at universities and industries. He served as an advisor for over 70 M.Sc. and Ph.D. theses and dissertations.

He has undertaken administrative positions at university, state, and federal institutions, including Head of Department, Director of Engineering Post-Grade Center, Vice-Rector for Research and Post Graduate Courses, Vice-Secretary for Science of the State of Rio de Janeiro, and Vice-Secretary for Higher Education of the Federal Ministry of Education. He was a coordinator of one of the branches of a science and technology program of the World Bank in Brazil. He is a fellow of the American Society for Metals and Materials and is a member of The Minerals, Metals & Materials Society. He has received several awards for technical and scientific works. He is currently president of the Brazilian Association for Metallurgy Materials and Mining (ABM) and is a consultant for the main Brazilian agencies for science and technology, CAPES, FAPERJ, and CNPq, where he is in the top national ranking (IA) as a researcher. He is coeditor of *Journal of Materials Research and Technology* and is a member of the editorial board of three international scientific journals.

About the Editors

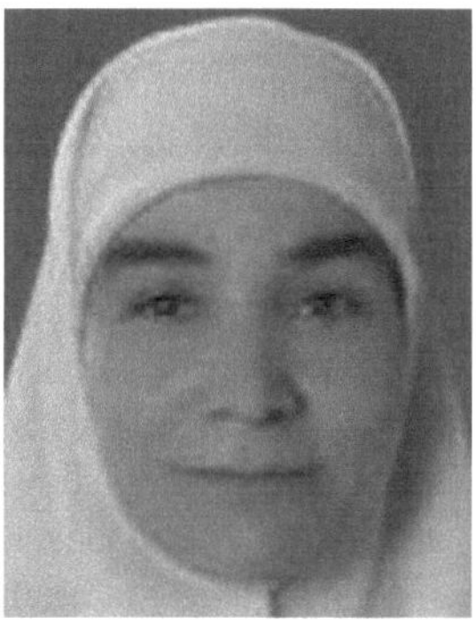

Shadia Ikhmayies received a B.Sc. and M.Sc. from the Physics Department at the University of Jordan in 1983 and 1987, respectively, and a Ph.D. on the topic of producing CdS/CdTe thin-film solar cells from the same university in 2002. She now works at Isra University in Jordan as an associate professor. Her research is focused on producing and characterizing semiconductor thin films, and thin-film CdS/CdTe solar cells. She also works in characterizing quartz in Jordan for the extraction of silicon for solar cells and characterizing different materials by computation. She has published 48 research papers in international scientific journals, 73 research papers in conference proceedings, and 3 chapters in books. She is the author of two books for Springer, *Silicon for Solar Cell Applications* and *Performance Optimization of CdS/CdTe Solar Cells* (both in production), editor of the book *Advances in II-VI Compounds Suitable for Solar Cell Applications* (Research Signpost), the book *Advances in Silicon Solar Cells* (Springer), an eBook series about material science (in development with Springer), and several TMS proceedings publications. She is the winner of the TMS Frank Crossley Diversity Award (2018) and the World Renewable Energy Congress 2018 (Wrec-18) Pioneering Award.

She is a member of The Minerals, Metals & Materials Society (TMS) and the World Renewable Energy Network (WREN). She is a member of the international organizing committee and the international scientific committee in the European Conference on Renewable Energy Systems (ECRES2015-ECRES2018). She is a member of the editorial board of the *International Journal of Materials and Chemistry* (Scientific & Academic Publishing) and has served as a technical advisor/subject editor for *JOM* (2014 and 2019). She has been a guest editor for topical collections from the European Conference on Renewable Energy Systems in the *Journal of Electronic Materials* and an editorial advisory board member for *Recent Patents on Materials Science* (Bentham Science). She is a reviewer for 24 international journals, was the Chair of the TMS Materials Characterization Committee (2016–2017), and has been lead organizer of more than four symposia at the TMS Annual Meeting and Exhibition.

Jian Li is a senior research scientist and program manager at CanmetMATERIALS in Natural Resources Canada. He obtained his B.Sc. in Mechanical Engineering from Beijing Polytechnic University, M.Sc. in Metallurgical Engineering from Technical University of Nova Scotia (TUNS), and Ph.D. in Materials and Metallurgical Engineering from Queen's University, Kingston, Ontario, Canada. He has broad experience in materials processing and characterization including alloys deformation, recrystallization, and micro-texture development. He has extensive experience in focused ion beam (FIB) microscope techniques. He is also an expert in various aspects of SEM-EDS and EPMA techniques. He holds a patent, has authored three chapters, and has published more than 140 papers in scientific journals and conference proceedings.

Carlos Mauricio Fontes Vieira graduated in Mechanical Engineering from Universidade Santa Úrsula (1991) and earned his master's (1997) and Ph.D. (2001) in Material Science and Engineering from Universidade Estadual do Norte Fluminense Darcy Ribeiro. He has tutored 11 doctors and 28 masters and published more than 160 articles in specialized journals and 250 in proceedings of congresses. He is an ad hoc consultant for national and international periodicals. He was Pro-Rector of Research and Post-Graduation at UENF. He is currently associate professor at the Universidade Estadual do Norte Fluminense Darcy Ribeiro. He has experience in Materials Science and Engineering, with emphasis on traditional ceramic materials and composite materials, working mainly in the following subjects: characterization of raw materials, body formulation, recycling of residues in clayey ceramics and composites materials incorporating residues and natural fibers in polymer.

Jean Igor Margem was born on April 4, 1956, in Campos dos Goitacazes, state of Rio de Janeiro, Brazil. He earned a Bachelor in Mechanical Engineering (1979) at the Gama Filho University, Rio de Janeiro, Brazil; a Master in Sanitary Engineering (1993) at the International Institute for Hydraulic, Delft, Netherlands; and a M.Sc. (2008) and Ph.D. (2013) in Materials Science and Engineering at the State University of the Northern Rio de Janeiro, Brazil. He was a professor of mechanical engineering at the Higher Teaching Institute (2008–2018), Campos dos Goytacazes, Brazil. He was the author of more than 30 papers and reports in the areas of sanitary engineering, sludge from industrial plant, and recycling and incorporation of wastes.

Fabio de Oliveira Braga graduated in Materials and Metallurgical Engineering at UENF/RJ/Brazil. He earned his Master and Doctor of Sciences (Materials Science) at the Military Institute of Engineering (IME/RJ/Brazil). Currently, he is a professor in the Faculty of the National Service of Industrial Apprenticeship (SENAI/RJ/Brazil) and at the Fluminense Federal University (UFF/RJ/Brazil), teaching topics such as materials science, extractive metallurgy, failure analysis, welding processes, and metallurgy. He is a coordinator in the field of metallurgy and materials in the oil and gas industry at SENAI (RJ/Brazil). His main research areas are ballistic protection, composites, natural fibers, multilayered armor systems, mechanical properties, and physical metallurgy.

Part I
Sustainable Ceramics

Study of Incorporation of Fuel and Fluxing Wastes in Red Ceramics

Gabriela Nunes Sales Barreto, Michelle Pereira Babisk, Geovana Carla Girondi Delaqua, Monica Castoldi Borlini Gadioli and Carlos Mauricio Fontes Vieira

Abstract Red ceramic industries are among the most important recyclers of industrial and urban wastes. The clays' heterogeneity allows the incorporation of several types of wastes, which can be classified as fuel wastes, fluxing wastes, and property-affecting wastes. The fuel wastes usually have a high amount of carbon-containing matter, that when heated cause exothermic reactions, releasing heat to the process. The fluxing wastes cause a reduction in the ceramic melting points. The main objective of this work was to study the behavior of ceramics incorporated with fuel wastes (*Eichhornia crassipes* dry biomass) and fluxing wastes (granite) as well as blends of both wastes. Different compositions were prepared with incorporation of different percentages of these wastes into red ceramics, shaped by uniaxial pressing and fired at 850 and 1050 °C. The technological properties tested were apparent dry density, linear shrinkage, water absorption, and flexural rupture strength. Results indicate that incorporation of these wastes into red ceramics is viable, which can significantly improve the evaluated properties of the ceramics, but the quantity and the firing temperature must be controlled.

Keywords Red ceramics · Fuel wastes · Fluxing wastes · *Eichhornia crassipes* · Granite

---

G. N. S. Barreto · M. P. Babisk (✉) · G. C. G. Delaqua · C. M. Fontes Vieira
Laboratory for Advanced Materials, State University of Northern Rio de Janeiro, Campos dos Goytacazes, RJ, Brazil
e-mail: michellebabisk@hotmail.com

G. N. S. Barreto
e-mail: gabibarreto93@gmail.com

G. C. G. Delaqua
e-mail: geovanagirondi@gmail.com

C. M. Fontes Vieira
e-mail: vieira@uenf.br

M. C. B. Gadioli
Núcleo Regional do Espírito Santo, Centro de Tecnologia Mineral, Cachoeiro de Itapemirim, ES, Brazil
e-mail: monicaborlini@gmail.com

S. Ikhmayies et al. (eds.), *Green Materials Engineering*, The Minerals, Metals & Materials Series, https://doi.org/10.1007/978-3-030-10383-5_1

Introduction

Human consumption habits combined with current industrial activities have generated an increasing amount of all waste types. Therefore, it is important to find practical and economic waste destination ways. The incorporation of industrial and urban wastes into red ceramic has been widely used nowadays, both in search for alternative raw materials and for an environmentally correct waste disposal.

The industry presents itself as an excellent alternative for different segments of waste management due to its large production, as well as traditional clay heterogeneity allows the incorporation of several types of wastes with a small sacrifice and, in some cases, improvement in the properties of the final product.

According to Vieira and Monteiro (2009), the different types of solid wastes that can be incorporated into red ceramics, according to their nature and to the ceramic products properties, can be classified in three categories:

- Fuel wastes—wastes that present high calorific value due to its high concentration of organic matter that contributes energetically to the ceramic sintering process through the energy release from the exothermic firing reactions.
- Fluxing wastes—wastes containing alkaline and earth alkaline compounds that in reaction with silica and alumina form liquid phases, increasing densification, reducing the melting point, helping to reduce the sintering temperatures.
- Property-affecting wastes—wastes containing substances that modify ceramic behavior such as those that interfere in mechanical resistance, water absorption, and linear shrinkage, but cannot be included in the first two categories [1].

Several works study the effects of these different wastes incorporated into red ceramics, but always separately: oil, paper, stone, glass, red mud, iron ore, biomass, different ashes and sludge [2–16]. However, few works mix two different waste types to analyze the effects of blend incorporation.

The objective of this work was to study the effect of incorporation of fuel (dry *Eichhornia Crassipes* biomass) and fluxing (granite) wastes in the properties of clay ceramics fired at different temperatures, as well as incorporated with blends of the two waste types with different proportions.

Materials and Methods

Materials

The raw materials used in this work were ceramic clay, dry macrophyte biomass waste, and granite waste. The ceramic clay was collected at ArtCerâmica Sardinha, located in the city Campos dos Goytacazes, in Rio de Janeiro, Brazil, and it is constituted by 1:1 portions of "weak" and "strong" yellow clays. The *Eichhornia crassipes* biomass was collected at the Lagoa do Vigário also located in Campos dos

Goytacazes. The granite waste was collected in a sawmill in the city of Cachoeiro de Itapemirim, Espírito Santo, from the filter press waste pile.

Preparation of Materials

The ceramic clay was dried at 110 °C for 24 h in a stove of the laboratory for Advanced Materials, LAMAV, of the State University of Northern Rio de Janeiro, Brazil. The dried ceramic clay was then separated in a jaw crusher, disintegrated in a porcelain mortar and to 42 mesh powders.

The biomass was washed in running water, dried in an air circulation stove and then in a laboratory stove at 60 °C. After dried, the material was ground in a knife mill and sieved through a 24-mesh sieve.

The granite waste was dried at room temperature for excessive moisture loss and dried in a laboratory stove at 110 °C, then sieved at 42 mesh.

Methods

Waste Characterization

Waste characterization was made to evaluate their behaviors as fuel and fluxing wastes. Differential scanning calorimetric (DSC) and thermogravimetric (TG) analysis were made in *Eichhornia crassipes* biomass to evaluate its thermal behavior and the reactions' nature. X-ray fluorescence (XRF) analysis was made in the granite waste to evaluate and quantify the presence of alkaline and earth alkaline compounds.

Preparation of Compositions

Table 1 presents the nine compositions formulated for the clay ceramic bodies incorporated with different percentages of wastes. The compositions were prepared with clay ceramic with additions of up to 5% biomass waste and up to 30% granite waste, as well as blend of both wastes.

Confection of Ceramic Bodies

The nine prepared composition masses were wetted with 8% water and conformed by uniaxial pressing in a hydraulic press with 15 TnF compression pressure, in a rectangular steel die (114 × 25 mm) to prepare the ceramic bodies. The samples were dried at room temperature and in a laboratory stove at 110 °C for 24 h, then fired at 850 and 1050 °C in a laboratory oven. The heating rate was 3 °C/min with

Table 1 Nomenclature and compositions (%weight)

Composition	Ceramic clay (%)	Dry biomass (%)	Granite waste (%)
A	100	0	0
B	80	0	20
C	70	0	30
D	97.5	2.5	0
E	95	5	0
F	77.5	2.5	20
G	67.5	2.5	30
H	75	5	20
I	65	5	30

an hour at the maximum temperature, with natural convection cooling after shutting off the oven.

Determination of Physical and Mechanical Properties

In order to evaluate the use of biomass as a raw material for the ceramic industry, the dry apparent density, linear shrinkage, and water absorption tests were done according to ASTM C373-72 [17] and the three-point flexural rupture strength, in accordance with ASTM C674-77 [18].

Results and Discussion

Figure 1 shows the DSC and TG biomass analysis, which performed to investigate its thermal behavior. In this figure, one should notice that around 100 °C there is a little endothermic peak that characterizes biomass loss of adsorbed water, associated with an average mass loss. Two well-defined exothermic peaks between 200 e 600 °C, associated with a ±55% mass loss, show that biomass releases heat in this temperature range.

This released heat will aid the sintering process of clay ceramics through heat releasing as well as energy savings during the firing process.

Table 2 shows the chemical composition of the granite waste, which is similar to that of clays used to make red ceramics. It is possible to observe the presence of SiO_2 (69.9%) followed by Al_2O_3 (17%), the significant amount of alkaline oxides, K_2O and Na_2O, in a total of 8.9% which act as fluxes and contribute to the formation of liquid phase and sintering of the ceramic part during firing [19].

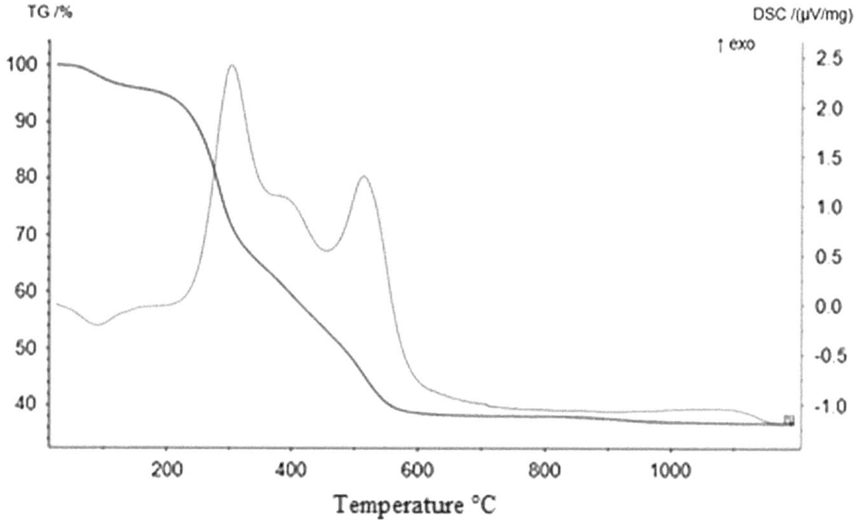

Fig. 1 DSC/TG *Eichhornia crassipes* dry biomass curves

Table 2 Chemical composition of granite waste (wt%)

Components	SiO_2	CaO	K_2O	Na_2O	Al_2O_3	Fe_2O_3	MgO	P_2O_5	TiO_2	LoI
	69.9	1.5	3.4	5.5	17	1.3	0.46	0.14	0.24	0.55

Another granite waste advantage is its low content of 1.3% Fe_2O_3, when compared to the clays used in Rio de Janeiro/RJ (Brazil) [20, 21]. The absence of iron compounds in the clay masses not only avoids damage to the wear of processing equipment, but also ceramic clays with a content below 3% Fe_2O_3 are indicated for the manufacture of light-colored products.

One final comment on the results of Table 2 is that the granite waste Low loss on ignition (LoI), 0.55%, indicates thermal stability. Excessive LoI can cause greater retraction and porosity after firing.

Figure 2 shows the dry apparent density of the nine studied compositions. The denser the ceramic body, the better "packaged" it is, because the closer its grains are, and there are less empty spaces between them. It is beneficial in the ceramic industry because the increase of particles contact area helps sintering. However, the increase in the densification causes a decrease in the ceramic body's permeability, which can impair the drying and the organic matter elimination during firing cycle. Statistically, one can observe that the granite and biomass waste incorporation did not change the dry apparent density when comparing to the pure ceramic clay.

Figure 3 shows the linear shrinkage of all compositions fired at 850 and 1050 °C. In this figure, one can observe that the linear shrinkage increases according to the increase of firing temperature. This is due to more efficient sintering at higher temper-

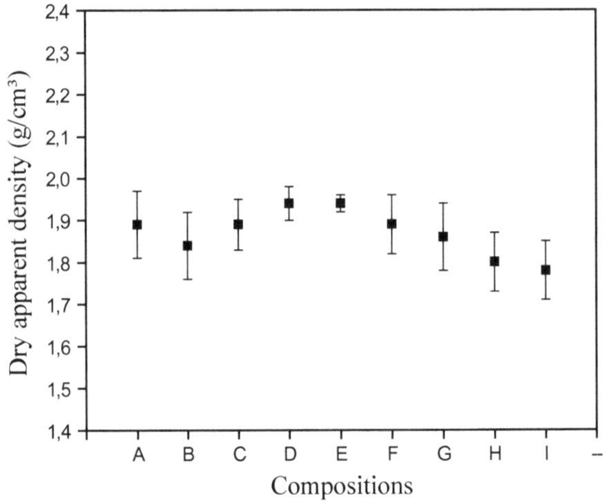

Fig. 2 Dry apparent density

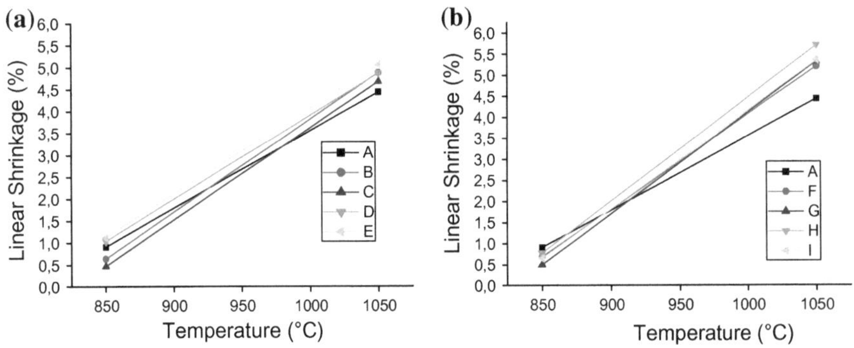

Fig. 3 Linear shrinkage of compositions: **a** A, B, C, D, E and **b** A, F, G, H, I

atures, which results in a larger pore closure and reduction of the bodies' dimensions [22].

One can also notice that, for both temperatures, the *Eichhornia crassipes* biomass incorporation (compositions D and E) increases the linear shrinkage of ceramic clay. Granite waste incorporation (compositions B and C), however, presents two different effects depending on the firing temperature. At 850 °C, granite waste causes a decrease in linear shrinkage, and at 1050 °C it causes an increase, because at this temperature the fluxes present in granite waste have already formed liquid phase, decreasing pore volume, and as a consequence, the structure contracts.

In blend compositions (F, G, H, and I), one can also notice two different effects depending on the temperature. At 850 °C, there was a decrease in linear shrinkage for all compositions, and this decrease is more pronounced in those with the highest amount of granite waste (G and I). At 1050 °C, there was an increase in linear

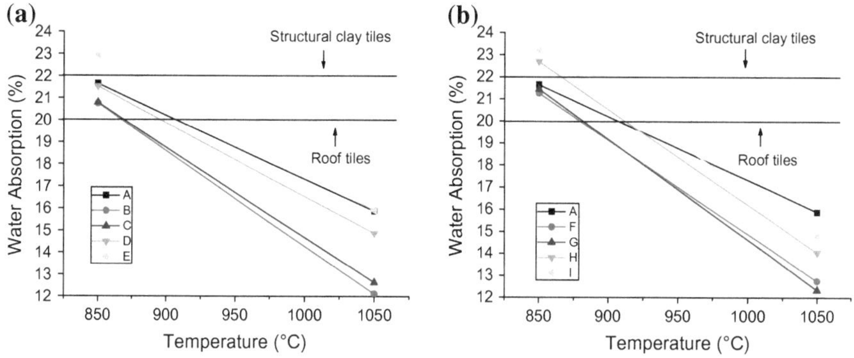

Fig. 4 Water absorption of compositions: **a** A, B, C, D, E and **b** A, F, G, H, I

shrinkage for all compositions due to its higher liquid phase formation because of the granite fluxing action, as discussed above.

Figure 4 shows the percentage water absorption of the compositions. In this figure, one can notice that at higher temperatures, water absorption decreases. This is because sintering is more effective at higher temperatures, resulting in a greater closure of open pores, thus reducing water absorption [22]. The lower water infiltration in the ceramic body determines, for example, the grater durability and resistance to the environment to which the material is exposed.

For the ceramics fired at both temperatures, the results indicated that the 20 and 30% granite waste incorporation (A and B) caused a decrease in water absorption comparing to the pure ceramic clay (A).

The water absorption increases with the *Eichhornia crassipes* biomass incorporation at 850 °C, including for blends, except for the composition D (2.5%), which maintains the same value of pure ceramic mass. These results are explained by the combustion of the biomass organic matter during the firing stage, associated with a mass loss that causes ceramic porosity, a natural behavior of this type of residue, as evidenced by thermal analysis. Meanwhile, at 1050 °C the formation of liquid phase helps to fill these pores and the water absorption values decrease.

It should be noted that the compositions with the lowest amounts of biomass cause a decrease in the water absorption and the ones with the highest amounts of biomass caused an increase.

For 1050 °C, all blend compositions caused a decrease in water absorption. The smaller the amount of biomass waste and the greater the amount of granite waste, the greater the water absorption reduction was, due to the liquid phase formation by the present fluxing oxides.

For structural clay tiles, according to NBR 15270 (2005), the water absorption index must be not less than 8% no more than 22% [23]. NBR 15310 (2009) indicates that the maximum permissible limit of water absorption for ceramic roof tiles is 20% [24]. The results obtained for all studied compositions, at 1050 °C, meet the normative limits for both manufactured pieces above. At 850 °C, the compositions

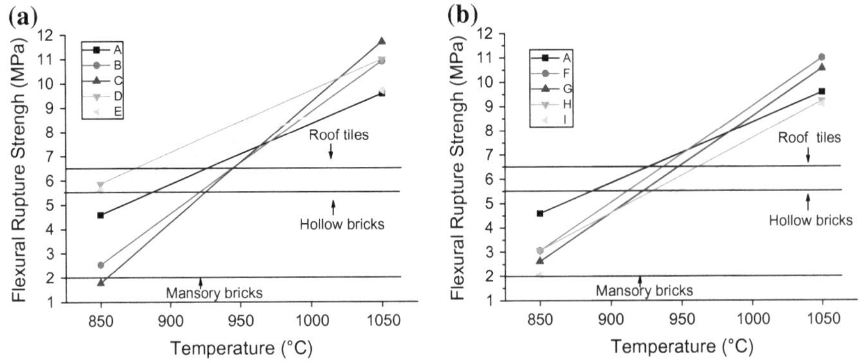

Fig. 5 Flexural rupture strength of compositions: **a** A, B, C, D, E and **b** A, F, G, H, I

are suitable for roof tiles manufacture, except all that incorporate 5% biomass (E, H, and I).

Figure 5 shows flexural rupture strength of the compositions. The flexural strength, or stress at which the material breaks, is the most important property for structural materials. One can observe that the flexural strength increases with the temperature increase, mainly to 1050 °C. As previously discussed, this is due to the sintering mechanisms that allow greater formation of liquid phase, thus reducing the porosity of the material and promoting a better consolidation of the particles.

In this graph, one can observe that in comparison with the pure ceramic (A), granite waste incorporation (compositions B and C) at 850 °C decreases the flexural rupture strength and *Eichhornia crassipes* biomass incorporation (compositions D and E) increases. The organic matter combustion of biomass releases heat that contributes to the ceramic sintering process, reaction that occurs between 200 and 600 °C as presented in thermal analysis, while the granite waste was inert at this temperature and therefore the presence of quartz and other minerals may have acted as stress concentrators. At 1050 °C, there was liquid phase formation, and the granite waste incorporation was effective.

All blend compositions fired at 850 °C, when compared to the pure ceramic (A), cause a decrease in the flexural rupture strength. At 1050 °C, all blend compositions with smaller biomass amounts (F and G) caused an increase.

According to Santos (1989), the minimum resistance required for the masonry bricks manufacture is 2 MPa, for hollow bricks is 5.5 MPa, and for roof tiles is 6.5 MPa [25]. The results obtained at 1050 °C for all compositions are above the minimum strength for manufacturing of these three products.

All studied compositions, at both temperatures, are above the minimum strength for masonry bricks manufacture, expect for the composition with 30% granite waste fired at 850 °C. It is noteworthy that in this temperature, all other compositions with granite waste incorporations (B, F, G, H, and I) only meet the requirements for masonry bricks manufacture and that the biomass incorporation (compositions D and E) fits the ceramic for hollow bricks manufacture.

Conclusions

Regarding to the objective of this work, one can notice that the *Eichhornia crassipes* biomass waste and granite waste own suitable characteristics for the use as raw materials in the ceramic industry. The characterization results, as well as its influences in ceramic properties, proved that they behaved as fuel and fluxing wastes, respectively, as proposed.

Thus, it is evident the direct influence of both wastes and their blends in the technological ceramic properties evaluated:

- Both the granite and biomass wastes as well as their blends did not case significant changes in the ceramic dry apparent density.
- Linear shrinkage increases for all ceramics incorporated with biomass waste. There were increasing and decreasing effects for ceramics incorporated with granite wastes and blends, with a decrease at 850 °C and an increase at 1050 °C.
- Water absorption increases with incorporation of 5% biomass waste, while granite waste incorporation caused a decrease. Blend compositions, except the ones with 5% biomass, caused a decrease in water absorption, for both temperatures.
- Flexural rupture strength, at the temperature of 850 °C, increased with biomass waste incorporation, while decreased with granite waste incorporation. At the temperature of 1050 °C, both wastes increased flexural rupture strength.

In addition to the improvements of some ceramic mechanical properties, the use of these wastes and their blends allows the recycling of a material that could be unduly disposed, causing environmental damage.

Acknowledgments The authors thank the Coordenação de Aperfeiçoamento de Pessoal de Nível Superior, CAPES, program PNPD20131134—31033016005P8 UENF/Materials Engineering and Science for the support of the research project.

References

1. Vieira CMF, Monteiro SN (2009) Incorporation of solid wastes in red ceramics: an updated review. Matéria 14(3):881–905
2. Ribeiro LS, Babisk MP, Prado US, Monteiro SN, Vieira CMF (2005) Incorporation of in natura and calcined Red Muds into clay ceramic. Mater Res 18(Suppl 2):279–282
3. Aguiar MC, Gadioli MCB, Babisk MP, Candido VS, Monteiro SN, Vieira CMF (2014) Clay ceramic incorporated with granite waste obtained from diamond multi-wire sawing technology. Mater Sci Forum 775–776:48–652
4. Babisk MP, Ribeiro WS, Aguiar MC, Monteiro SN, Vieira CMF, Vidal FWH (2015) Characterization of a quartzite residue and its application in red clay ceramics. Mater Sci Forum 805:41–546
5. Amaral LF, Carvalho JGR, Silva BM, Delaqua GCG, Monteiro SN, Vieira CMF (2018) Development of ceramic paver with ornamental rock waste. J Mater Res Technol JMR&T 2018:01–10
6. Delaqua GCG, Vieira CMF, Amaral LF, Monteiro SN (2016) Incorporation of dry biomass of *Salvinia Auriculata* aubl. In: The process of Phytoremediation in Ceramic Productio. 60° Congresso Brasileiro de Cerâmica, pp. 491–490. (in portuguese)

7. Morais ASC, Vieira CMF, Sánchez RJ, Monteiro SN, Candido VS, Ferreira CL (2016) Fluorescent lamp glass waste incorporation into clay ceramic: a perfect solution. JOM 68:2214–2225
8. Babisk MP, Altoé TP, Lopes HJO, do Prado US, Gadioli MCB, S.N. Monteiro SN, Vieira CMF (2014) Properties of clay ceramic incorporated with red mud. Mate Sci Forum, 798–799:509–513
9. Borlini MC, Vieira CMF, Monteiro SN (2006) Incorporation of ash from sugar cane bagasse into clay bricks. Ind Ceram Faenza 26:23–29
10. Babisk MP, Oliveira CN, Soares MP, Vieira CMF, Monteiro SN, Alexandre J (2014) Effect of banana fiber in the properties of clayey ceramic. Mater Sci Forum (Online) 798–799:229–234
11. Monteiro SN, Alexandre J, Margem JI, Sánchez RJ, Vieira CMF (2008) Incorporation of sludge from water treatment plant into red ceramic. Constr Build Mater 22:1281–1287
12. Pinheiro RM, Vieira CMF, Sánchez RJ, Monteiro SN (2008) Recycling of waste from the paper production into red ceramic. Matéria (UFRJ). 13:220/10987-227. (in portuguese)
13. Souza CC, Vieira CMF, Monteiro SN (2008) Microstructural changes of clayey ceramic incorporated with iron ore tailings. Matéria (UFRJ). 13:194/10982-202. (in portuguese)
14. Monteiro SN, Vieira CMF, Ribeiro MM, Silva FAN (2007) Red ceramic industrial products incorporated with oily wastes. Constr Build Mater Scotland 21:2007–2011
15. Vieira CMF, Teixeira SS, Monteiro SN (2009) Effect of firing temperature on the properties and microstructure of red ceramic incorporated with grog. Cerâmica (São Paulo. Printed) 55:332–336. (in portuguese)
16. Coutinho NC, Vieira CMF (2016) Characterization and incorporation of MSWI ash in red ceramic. Cerâmica 62:249–255 (in portuguese)
17. Vieira CMF, Monteiro SN (2006) Clayey ceramic incorporated with oily waste from the petroleum industry. J Materia 11(3):217–222 (in portuguese)
18. ASTM—American society for testing and materials (1972) Water Absorptin, Bulk Density, Apparent Porosity and Apparent Specific Gravity of Fired Whiteware Products, C373-72, USA
19. ASTM—American society for testing and materials (1977) Flexural Properties of Ceramic Whiteware Materials, C674–77, USA
20. Monteiro SN, Vieira CMF (2002) Tile & Bricks Int. Characterization of clays from Campos dos Goytacazes, north Rio de Janeiro State (Brazil) 18:152–157
21. Babisk MP, Ribeiro AP, Monteiro SN, Vieira CMF (2015) Development of ceramics based on clays from different regions in the state of rio de janeiro. Brazil. Mater Sci Forum 805:530–535
22. Pereira AC, Monteiro SN, Assis FS, Braga FO, Vieira CMF, B.M. Silva BM (2017) Development of clay ceramics incorporated with ornamental stone wastes for pavement application. ABM Week. 72(1):429–440. (in portuguese)
23. ABNT—Brazilian Standars Associaton (2005) NBR 15270. Ceramic components Part 1: Hollow ceramic blocks for non load-bearing masonry—Terminology and requirements. (in portuguese)
24. ABNT—Brazilian Standars Associaton (2009) NBR 15310 Ceramic componentes—Ceramic roof tiles—Terminology, requirementes and testing methods. (in portuguese)
25. Santos, PS (2ed) (1989) Clay science and technology. Edgard Blücher, São Paulo, SP (in portuguese)

Mechanical and Thermal Behaviour of Clay Filled Recycled Low Density Polyethylene

G. C. Onuegbu, G. O. Onyedika and M. O. C. Ogwuegbu

Abstract The mechanical and thermal properties of clay powder filled recycled low density polyethylene (rLDPE) have been determined. The filler loadings considered were at 0, 5, 10, 15 and 20 wt% and particle sizes of 90 and 425 μm. The incorporation of clay powder into rLDPE resulted in improvement of the tensile strength, tensile modulus, fatique, hardness, and impact strength of the composites. The results also show that the mechanical properties increase with increases in filler clay contents and decrease in particle sizes. The strain at break and shear strength were found to decrease with increases in clay loadings. The thermo-graphical analysis on the rLDPE/clay composite showed that increase in filler loading decreases the weight lost thereby enhancing the thermal stability of the rLDPE composite.

Keywords Mechanical properties · Thermal analysis · Composite · Recycled low density polyethylene

Introduction

Composite materials are experiencing an increase in interest due to their high demands in industry. These are evaluated according to the materials' good specific properties performance which depends on the characteristics of each of their components, the arrangement of the reinforcement in the matrix, and interface quality. In recent years, interest has grown worldwide in the incorporation of cellulosic fibers and clay hybrid fillers in composites, because present studies show that for their good mechanical properties, the fiber and clay can act as reinforcing polymeric matrices in addition to environmental and sustainability issues [1–3, 7].

G. C. Onuegbu (✉)
Department of Polymer and Textile Engineering,
Federal University of Technology, Owerri, Nigeria
e-mail: onuegbugc@gmail.com; gc.onuegbu@yahoo.com

G. O. Onyedika · M. O. C. Ogwuegbu
Department of Chemistry, Federal University of Technology, Owerri, Nigeria

S. Ikhmayies et al. (eds.), *Green Materials Engineering*, The Minerals, Metals & Materials Series, https://doi.org/10.1007/978-3-030-10383-5_2

Among the various types of polymer-based composite, thermoplastic-based composites are gaining interest from both academia and industry because of their unique ability to be recycled and reclaimed [5]. The low density polyethylene (LDPE) can be used as food packaging materials, beer bottles, carbonated drinks, juice bottles and thermoformed containers, and pharmaceutical packaging. Recycled polymers such as LDPE have poor properties compared to virgin polymers and are used in application requiring low strength materials [4, 8, 9]. In this study, an attempt is made to improve the properties of the recycled polymer low density polyethylene (rLDPE) by adding clay. Recycled LDPE filled with clay composite were prepared, tested, and evaluated for the effect of reinforcement concentration on the mechanical and thermal properties.

Materials and Methods

The recycled low density polyethylene (rLDPE) was obtained as a factory waste from Le- Winners Industries Internationale, Owerri. The company obtained their printed low density polyethylene (LDPE) from CEE-PLAST industry, Osisioma Ngwa Aba, Abia state, Nigeria. The LDPE is a product of INDORAMA group, a subsidiary of Eleme Petrochemical Company Limited River State, Nigeria. It has a melt flow index of 0.4 g/10 min at 190 °C and density of 0.922 g/cm^3. The LDPE was extruded into films and printed before use in the sachet water packaging. The used sachet water (recycled LDPE) was cut into pieces and drie at a temperature of 30 °C. The dried rLDPE was then crushed into fine particles.

The reinforcing agent in this study was the run off mined clay. The clay was obtained from Awo, Oru West Local Government Area of Imo State, dried, and crushed. The crushed banana fibers were sieved and paricle sizes of 90 and 425 μm was selected for this study.

Composites Production

The run-off mined clay was collected from local miners, sun dried, pulverized in the grinding machine and subjected to sieve analysis which was used to classify it into different particle sizes. Particle sizes such as 90 and 425 μm were obtained. The recycled low density polyethylene/clay composite at various particle sizes of 90 and 425 μm was prepared by thorough mixing different quantities of rLDPE/clay to obtain 5, 10, 15, and 20% clay loading. The recycled low density polyethylene and the sieved clay were melted in an Injection moulding machine at 173 °C and the resultant composites were produced as sheets. At the time of curing, a compressive pressure of 100 atm was applied. Replicated samples of clay reinforced polyethylene matrix were afterwards subjected to hardness, flexural tensile, shear and fatigue tests and other test.

Results and Discussion

Mechanical Properties

The mechanical properties of rLDPE and clay/rLDPE composites have been determined and are discussed below.

Tensile Strength

Figure 1 shows the effect of clay loading on the tensile strength of the recycled low density polyethylene composite. The tensile strength of unfilled rLDPE is 700 mpa and the filled samples showed greater tensile strength than the unmodified sample except at 5 and 15% of clay (filler) loading at 425 μm particle size which was later increased. At 90 μm particle size, the tensile strength increased with increases in filler loading. The increase in tensile strength may be attributed to the strengthening effect of the clay incorporated into the polymer matrix.

Figure 1 also shows that the tensile strength decreased with increasing particle size. This is because the smaller particle sizes provide larger surface area for interaction of filler with the matrix thereby increasing the ability of the reinforcers to support stress transferred from the matrix.

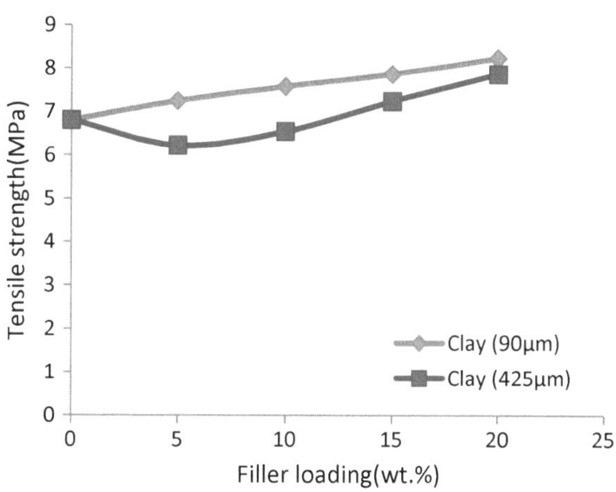

Fig. 1 Plot of tensile strength versus filler loading for Clay/rLDPE composite at two particle sizes

Fig. 2 Strian at break versus filler loading for rLDPE/clay composite at two particle sizes

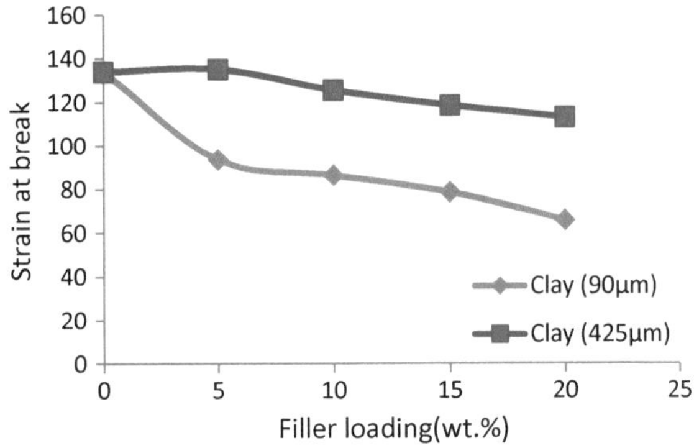

Strain at Break

The strain at break of the unfilled is 143 and rLDPE/clay composites decrease with increase in filler content for all the filler loadings for both particle sizes (Fig. 2). The strain at break of rLDPE/clay at 425 µm composites was observed to be higher than that at 90 µm rLDPE/clay. The strain at break of natural filler/polymer composites has been observed to reduce as the fibres' stiffness increases.

Tensile Modulus

Figure 3 show that the tensile modulus of unfilled rLDPE is 2 Mpa. It shows that the modulus of the composites increased with increase in filler loading. The observation highlights the fact that the incorporation of fillers into the polymer matrix improved the stiffness of the composites [6]. On the other hand, smaller particles enhance better dispersion within the matrix than larger particles.

Fig. 3 Tensile modulus versus filler loading for rLDPE/clay composite at two particle sizes

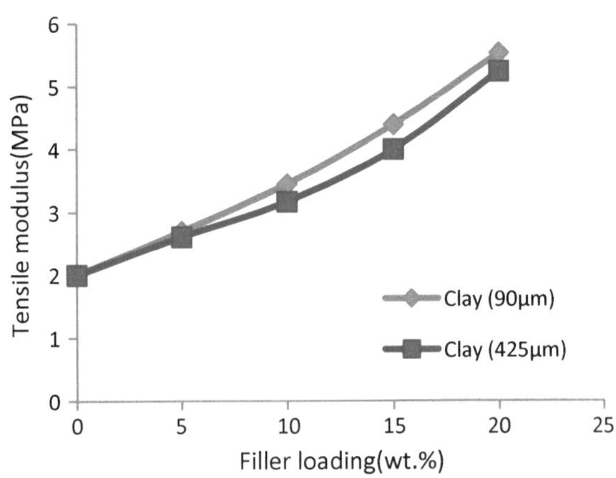

Fig. 4 Flexural strngth versus filler loading for rLDPE/clay composite at two particle sizes

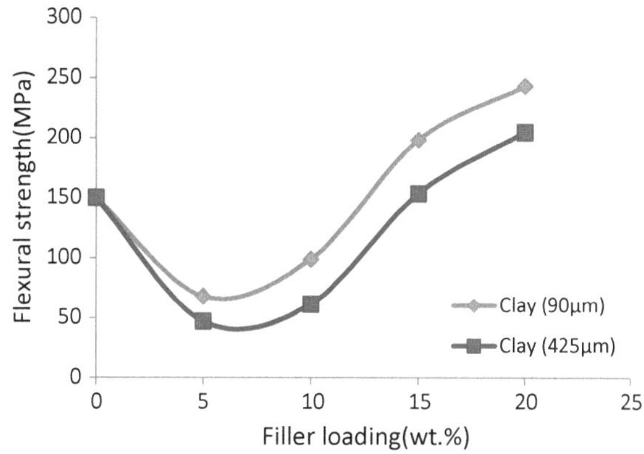

Fig. 5 Shear strength versus filler loading for rLDPE/clay composite at two particle sizes

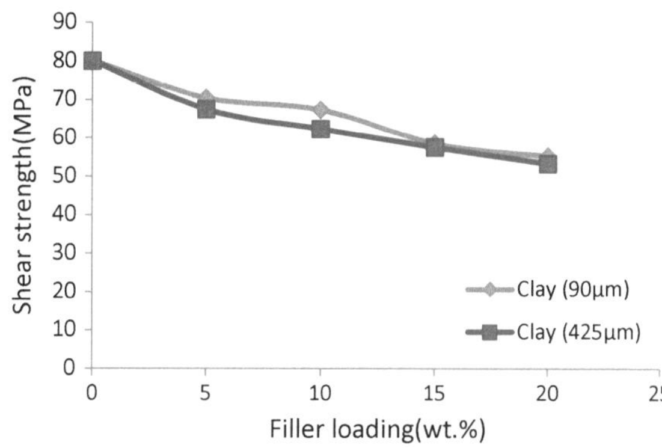

Flexural Strength

From Fig. 4, the flexural strength of unfilled rLDPE is 150 Mpa. The flexural strength of the composite at the two particles sizes considered decreased initially with increases in filler loading and later increased from 10 wt% of clay. This suggests that initially there was not much stress transfer from matrix to the filler for the two particle sizes considered. This shows that incorporation of more reinforcing agents has the ability to resist the bending forces applied. The figure also shows that flexural strength at 90 μm is higher than that of 425 μm.

Shear Strength

Shear strength is the amount of stress required to pierce or cut across a given area of a material. The shear strength of rLDPE/clay is shown in Fig. 5. The shear strength of unfilled rLDPE is 80 Mpa. The figure shows that the shear strength decreased with increase in filler loadings and increases in particle sizes.

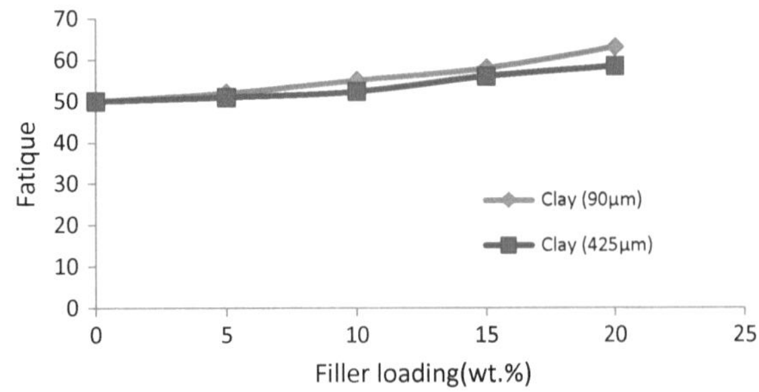

Fig. 6 Fatique versus filler loading for rLDPE/clay composite at two particle sizes

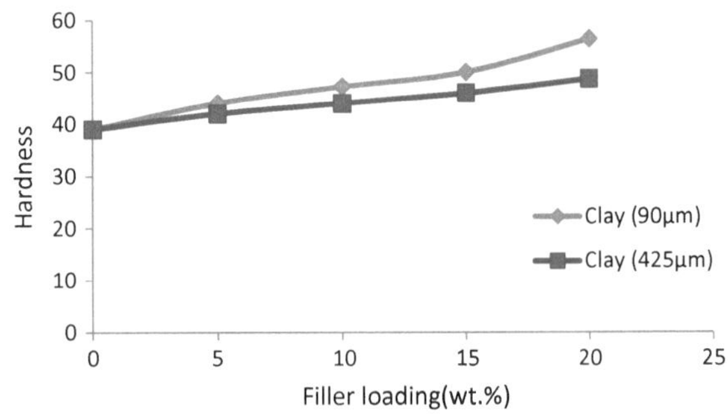

Fig. 7 Hardness versus filler loading for rLDPE/clay composite at two particle sizes

Fatigue Strength

Fatigue strength is the condition whereby a material cracks or fails as a result of repeated (cyclic) stresses applied below its ultimate tensile strength. Figure 6 show that the fatigue strength of unfilled rLDPE is 50 Mpa. The figure shows that the fatigue strength increased with increases in filler loadings and decrease in particle sizes.

Hardness

The hardness of clay/RLDPE is shown in Fig. 7. The hardness of unfilled RLDPE is 39. The clay/RLDPE composites exhibited higher hardness than the unfilled RLDPE. The figures shows that the hardness increased with increases in amount of clay incorporated into the matrix. It has been reported that the incorporation of fillers into polymer matrices makes the resulting composites harder [6]. The hardness decreased with increases in particles sizes.

Table 1 TGA analysis of the decomposition temperature and weight loss in rLDPE/clay composite

Materials	T_{onset} (°C)	T_{max} (°C)	Weight loss (%)
Neat rLDPE	238.12	378.91	98.52
rLDPE + 5wt% Awo Clay	437.86	597.78	96.29
rLDPE + 10 wt% Awo Clay	445.98	597.45	95.55
rLDPE + 15 wt% Awo Clay	441.91	596.81	93.96
rLDPE + 20 wt% Awo Clay	451.84	596.79	90.98

Thermogravimetric Analysis (TGA) and Differential Scanning Calorimetry (DSC) Results

The TGA and DSC analysis was performed on the rLDPE/clay composite at 90 μm particle size as shown in Table 1. The selection for this sample was based on the good results obtained from its mechanical properties. This, however, necessitate the investigation of the thermal behaviour of the selected composite materials. TGA analyses were performed on neat rLDPE and rLDPE/clay composites at 90 μm particle sizes with varying clay loading. From Table 1, the weight loss recorded by the addition of 5, 10, 15, and 20 wt% clay into the rLDPE were 96, 95, 94, and 91% respectively. These observations suggest that the increase in the amount of clay filler enhances the thermal stability of the rLDPE material.

Figure 8 shows the thermographs for the pure rLDPE and the composites. The neat rLDPE has a maximum melting temperature (T_{max}) of 102 °C and corresponding melting enthalpy (ΔH_m, Jg^{-1}) of −78.74 with thermal stability at 125 °C. For rLDPE + 5 wt% clay, rLDPE + 10 wt% clay, rLDPE + 15 wt% clay, and rLDPE + 20 wt% clay composites, their initial melting temperatures were at 107 °C. The enthalpies ΔH_m, were −90.12, −103.86, and −101. 78 Jg^{-1} respectively. The highest stability temperature was seen for the rLDPE + 15 wt% clay composite and thermally stable at 133 °C.

From the TGA and DSC analysis, the temperature of the extruder should be set within 108 °C to facilitate the melting of the matrix to avoid time wastage of composite material in the extruder Clay has been utilized successfully in preparing recycled low density polyethylene composites. The tensile strength and module, fatigue, hardness, and impact strength of the recycled low density polyethylene composites where found to increase with increase in filler loading and decrease in filler particle size, the strain at break and shear strength of the prepared composites decreased with increase in filler loadings and particle sizes. However the flexural strength of the recycled low density polyethylene filled clay decreased initially with increase in filler loadings and later increased from 10wt% of clay. The thermal analysis showed that increase

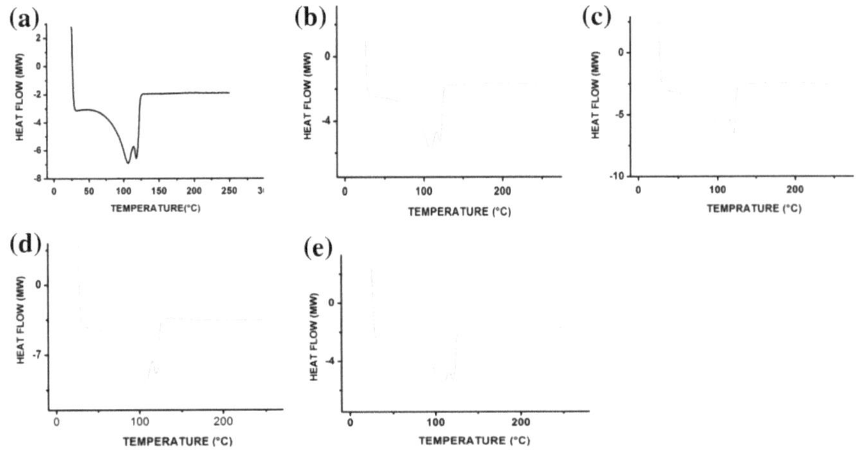

Fig. 8 The Thermographs of rLDPE and clay/rLDPE (**a** is rLDPE + 0 wt% clay; **b** is rLDPE + 5 wt% clay; **c** is rLDPE + 10 wt% clay; **d** is is rLDPE + 15 wt% clay; **e** is rLDPE + 20 wt% clay)

in filler loading reduced the weight lost thereby improving the thermal stability of composite. The use of clay as a filler has highlighted the importance of using clay as filler for recycled low density polyethylene.

Conclusion

Clay has been utilized successfully in preparing recycled low density polyethylene composites. The tensile strength, tensile modulus, fatique, hardness, and impact strength of the recycled low density polyethylene composites were found to increase with increase in filler loading and decreases in filler particle size. The strain at break and shear strength of the clay/rLDPE decreased with increase in filler loadings and particle sizes. However, the flexural strength of the rLDPE composite decreased initially with increase in filler loading and later increased. The thermal analysis showed that increase in filler loading reduced the weight lost thereby improving the thermal stability of the composite. The use of clay as a filler has highlighted the importance of using clay as filler in composite formulation.

References

1. Bisanda ETN (2000) The effect of alkali treatment on the adhesion characteristics of sisal fibres. Appl Compos Mater 7:331
2. Doh G-H, Lee SY, Kang IA, Kong YT (2005) Thermal behaviour of liquefied wood polymer composites. Compos Struct 68:103–108

3. Girisha C, Sanjeevamurthy, Guntiranga S (2012) Effect of alkali treatment, fiber loading and hybridization on tensile properties of sisal fiber, banana empty fruit bunch fiber and bamboo fiber reinforced thermo-set composites. Int J Eng Sci Adv Technol 2:706–711
4. Liu H, Qingxiang Wu, Zhang Q (2009) Preparation and properties of banana fiber-reinforced composites based on high density polyethylene (HDPE)/Nylon-6 blends. J Bioresour Technol 100:6088–6097
5. Murthy NN (2009) Development of commercial applications for recycled plastics using finite element analysis. M.S.C thesis, Brigham Young University
6. Onuegbu GC, Obidiegwu MU, Onyedika GO (2016) Matlab analysis of fleural strength and modulus of hybrid nano polyster composite. FUTO J Ser 2(1):96–106
7. Ramya JR (2010) Preparation, characterization and properties of injection molded grapheme nanocomposites. M.S.C thesis, Wichite State University, Kansas, USA
8. Tsai SW, Pagano NJ (2011) Invariant properties of composite materials. In: Tsai SW, Halpin JC, Pagano NJ (eds) composite materials workshop. Technomic Publishing Co. Inc., Stamford, 233–253
9. Utracki LA (2003) Introduction to polymer blends. In: Utracki LA (ed) Polymer Blends Handbook, vol 1. Kluwer Academic Publishers, Dordrecht, pp 3–6

Physical and Mechanical Properties of Artificial Stone Produced with Granite Waste and Vegetable Polyurethane

Maria Luiza P. M. Gomes, Elaine A. S. Carvalho, Larissa N. Sobrinho, Sergio N. Monteiro, Rubén J. S. Rodriguez and Carlos Mauricio Fontes Vieira

Abstract The use of ornamental stones waste incorporated in resin to produce artificial stone has been an interesting alternative to reduce the increasing amount of discarded residue in the environment; in addition, it is technically and economically viable. This research aims at producing artificial stone from 85 wt% of granite waste incorporated in 15 wt% of vegetable polyurethane to be used in the construction industry. Initially, the best particle size composition of the residue was determined in order to better pack the particles. The vacuum vibro compression was used for the production of artificial stone plates. The study evaluated the mechanical and physical properties of the produced artificial stone as well as of the natural ornamental stone, granite. The artificial material using granite showed superior properties when compared to natural granite. The values of density, water absorption, and apparent porosity were, 2.24 g/cm^3, 0.19%, 0.42%, respectively. The flexural strength obtained by the rupture stress was 18 MPa. The results in tests are within the standard expected range, and show the good performance of the artificial stone.

Keywords Artificial stone · Residue · Granite · Vegetable polyurethane

Introduction

Brazil is one of the largest producers of ornamental rocks. The Brazilian production of natural rocky materials for ornamentation and coating was estimated, by ABIROCHAS, in 9.2 Mt in the year 2017, with granite corresponding to 55% of the total produced. These materials are in their vast majority used in construction [1].

M. L. P. M. Gomes (✉) · E. A. S. Carvalho · L. N. Sobrinho · R. J. S. Rodriguez · C. M. Fontes Vieira
Advanced Materials Laboratory – LAMAV, State University of the Northern Rio de Janeiro - UENF, Av. Alberto Lamego, 2000, Campos dos Goytacazes 28013-602, Brazil
e-mail: malu_pmg@hotmail.com

S. N. Monteiro
Materials Science Department, Military Institute of Engineering - IME, Praça General Tibúrcio, 80, Urca, Rio de Janeiro, RJ 22290-270, Brazil

S. Ikhmayies et al. (eds.), *Green Materials Engineering*, The Minerals, Metals & Materials Series, https://doi.org/10.1007/978-3-030-10383-5_3

However, during the stages of the production process of ornamental rocks since the extraction until obtaining the final product, a large amount of waste is generated. The extraction step generates an amount of byproduct between 40% and 60% of the global production, and another 30–35% is generated as a result of cutting and lapping [2]. The vast amount of waste has been currently deposited in inappropriate disposals such as the ocean, river, and lakes causing environmental problems. With a view to remove harmful residues from the environment, currently researchers have tested the viability of valorizing ornamental stone waste as the raw material in new sustainable materials [3, 4]

At the same time, the use of polymers in various types of products has been growing over the years generating a large amount of waste, which is too often discarded improperly. This fact is getting worse once that kind of material takes years to degrade, increasing the impacts caused by them. The use of biodegradable polymers stands as an alternative to minimize this problem, once their technical and economic feasibility has great potential for expansion [5].

In this context, in order to reduce environmental impacts and meet requirements imposed by environmental laws, one of the alternatives found is to reuse and recycle this type of waste for the production of artificial rocks, a type of polymer composite [6–8]. The industrialized stone consists of a high content of natural aggregates (until 95%) agglomerated by polymer resin. The aggregates that form artificial stone may be composed of quartz sand, marble, glass crystals, and others [8]

The production of artificial rocks is a worldwide lucrative activity and has been growing over the years. The demand in the US Market was of 6.7 million m^2 in 2016, representing a growth compared to 2011 [9]

The artificial stones, even though they are more expensive than the natural ornamental stones, they are very well accepted by consumers due to their technical quality when compared to natural rocks. Artificial rock is lighter than the natural one due to low density of the polymer used as the matrix. Another advantage of the artificial rock is the low amount of pores and voids found, while the high amount of pores in the natural ornamental stones provides them micro-structural defects causing fluid to be absorbed more easily. Also, the pores act as stress concentrators causing the cracks to spread more easily. Just this fact itself gives the natural ornamental stones inferior mechanical properties, so that the lifespan of the artificial rocks becomes longer [10, 11].

Many researches are found for the production of artificial rock primarily using the epoxy resin and polyester as matrix. The present work aims at developing an artificial rock from the granite recycling, in addition to using a biodegradable resin from castor oil, producing a sustainable material as well as technically and economically feasible.

Materials and Methods

In this study, it was used granite residue supplied by the Brumagran Company, Brazil and the polymer used as matrix was the vegetable polyurethane from castor

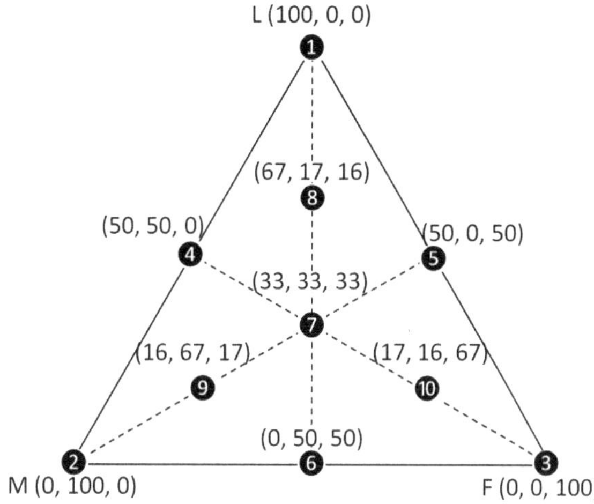

Fig. 1 Ternary diagram with the ten mixtures based on the complete cubic model of the simplex. amounts (wt%) of large (L), medium (M), and fine (F) particles (Carvalho et al. 2018)

oil supplied by Imperveg, Brazil. The residue was sieved according to ABNT NBR 7181 [12] and classified in three granulometric ranges: large (from 2 mm up to 0.42 mm), medium (from 0.42 mm up to 0.075 mm), and fine particles (with grains with size inferior to 0.075 mm).

Based on these three granulometric classifications, ten distinct mixtures were proposed for the residue for best-packed condition using the simplex-lattice design (SLD) numerical modeling methodology. Figure 1 shows schematically the mixture compositions investigated, which is similar to that applied elsewhere [13].

The determination of best-packed composition for the granite was associated with the highest dry apparent density. This density was obtained as per the Brazilian NBR 3388 standard [14] for ten different compositions considered in the simplex method, Fig. 1. For each composition, the test was made three times to assure statistical validation. Each sample of granite compositional mixture was placed in a steel vessel and allowed to vibrate for 2 min under a load of 10 kg. The mixture was weighed, and the apparent density was calculated. Composition with highest apparent bulk density, associated with best-packed particles, was selected to produce the artificial stone.

The artificial stone plates were fabricated by mixing with 85 wt% of granite and 15 wt% of still fluid vegetable polyurethane resin inside a 100 × 100 × 10 mm^3 steel mold. Initially, the residue was stove dried at 100 °C for 24 h to release moisture and then kept inside a desiccator jar until returning to room temperature (RT), around 25 °C. Each artificial stone using granite (AOG) fabrication was carried out under 600 mmHg vacuum, while the mold was vibrated for 2 min under a pressure of 10 MPa. The last stage of fabrication consisted of curing at 90 °C for 20 min. The artificial stone produced and the natural granite were characterized in order to compare their properties and to verify the efficiency of the material developed.

The density, water absorption, and apparent porosity of the AOG and granite were determined as per the Brazilian NBR 15845 standard [15]. Six prismatic specimens, cut from the AOG plate, granite with dimensions of $100 \times 25 \times 10$ mm^3, were three points bend tested in a model 5582 Instron machine following the recommendation for agglomerated stones as per the Spanish EN 14617 standard [16] as well as the Annex F of the Brazilian NBR 15845 standard [15].

Results and Discussion

Table 1 presents the values obtained by the SLD method for the average density of vibrated mixtures of granite according to Fig. 1. In this table, the highest density value was 1.951 g/cm^3, corresponding in Fig. 1 to the mixture containing 67% of large particles (L), 17% of medium particles (M), and 16% of fine particles (F) of granite. This is considered to be associated with the best-packed mixture and was used in the production of artificial stone.

Table 2 presents the experimental results found for: density, water absorption and porosity of the granite and AOG. The density value for the granite was 2.62 ± 0.01 g/cm^3 and for AOG was 2.24 ± 0.01 g/cm^3. The lower value of the apparent porosity of the AOG can be explained by the use of polymer and a low density material. Lee et al., in his work of artificial stone using glass fiber/PET has varied vacuum, vibration conditions, and compression level in the production process and found density values within 2.03–2.45 g/cm^3 [17]. The density value reached for AOG in this research is within the values found by the author.

Table 1 Vacuum and vibrate density of granite

Mixture	Average density (g/cm^3)
1	$1.493 \pm 0,041$
2	1.778 ± 0.017
3	1.600 ± 0.079
4	1.812 ± 0.032
5	1.913 ± 0.037
6	1.813 ± 0.060
7	1.923 ± 0.087
8	1.951 ± 0.018
9	1.906 ± 0.013
10	1.678 ± 0.019

Table 2 Physical properties of artificial stone

Physical properties	Granite	AOG
Density (g/cm^3)	2.62 ± 0.01	2.24 ± 0.01
Warter absorption (%)	0.38 ± 0.02	0.19 ± 0.02
Apparent porosity (%)	0.99 ± 0.06	0.42 ± 0.05

Table 3 Flexural strengh values for the granite and AOG

Material	Flexural strengh (MPa)	
	Granite	AOG
CP1	12.18	19.96
CP2	12.20	20.00
CP3	14.31	16.69
CP4	14.99	16.26
CP5	13.39	17.39
CP6	13.91	18.29
Average	13.50 ± 1.14	18.10 ± 1.62

The water absorption value of granite and artificial stone was 0.38 ± 0.02% and 0,19 ± 0.02%, respectively. The artificial stone value is 50% lower than the granite one. The values for water absorption reported by artificial marble manufacturers are in the range from 0.09 to 0.40%, so the presented value for water absorption is within the aforementioned industrial range [18]. Carvalho et al. found 0.25% for artificial stone using epoxy and quarry dust [13]. With these comparisons, it is possible to indicate the good result obtained for AOG, meaning that the material can be applied in humid environments.

In regard to apparent porosity, the values found for granite and artificial stone were 0.99 ± 0.06% and 0.42 ± 0.05%, respectively. The water absorption value of artificial stone was approximately 40% below the found for the ornamental one. Chiodi and Rodriguez classified the materials used to covering in construction and determined that a value below 0.5% represents that the stone has a very high quality, which represents the satisfactory value found for the AOG [19]. It can indicate the good adhesion of the resin to the granite load avoiding the emergence of voids.

Table 3 presents the flexural strength values for the granite and AOG. Figure 2 shows the flexural stress versus strain curves of the granite, together with that for the AOG. In the bending flexural strength test at three points it was obtained the value of 13.50 ± 1.14 MPa for granite and 18.10 ± 1.62 MPa for the artificial stone. As expected, the granite rupture strain was lower because it presents higher level of porosity as proven in the test of physical properties.

The artificial stone presented higher mechanical properties when compared to natural stone, it can be explained due to the molecular interconnections between the matrix and the load. The low dispersion results indicate a considerable mechanical stability of the artificial material produced.

Other producers of industrial stone reported values within the range of 13.6–17.2 MPa [8]. Borselino et al. produced an artificial marble with epoxy resin and obtained a result values of flexural strength between 10.6 and 22.2 MPa that is, the result obtained in this study is within the expected results according to the commercials manufacturers and researched stones [20].

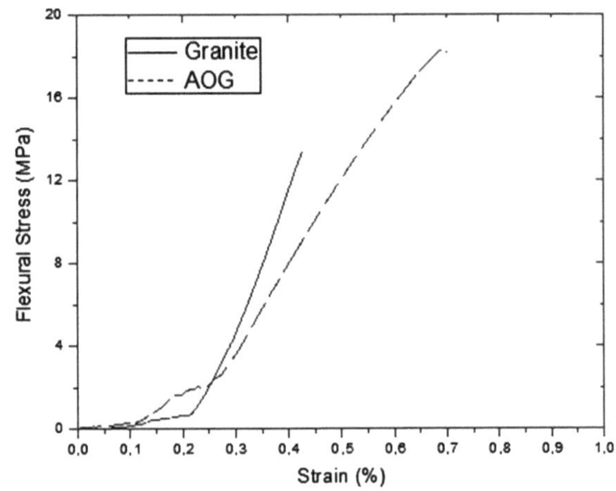

Fig. 2 Flexural stress x strain curves for neat epoxy and AOG

Conclusions

- A novel artificial stone (AOS) was for the first time produced with granite residue and vegetable polyurethane.
- The granite particle mixture was determined based on the highest density in terms of the best-packed characterization, found by means of the simplex-lattice design method.
- The artificial stone produced with granite obtained superior properties than the natural ornamental granite.
- The values of density, 2.24 g/cm^3, water absorption, 0.19%, and apparent porosity, 0.42% for the AOG are within the Brazilian standards and considered adequate when compared to other ornamentals stones used in civil construction.
- The flexural strength of 18 MPa for the AOG was considered adequate when compared to other ornamental stones used in civil construction.
- Through the obtained results in the evaluation of the physical and mechanical properties, the good adhesion of the granite particles in the matrix can be verified.

Acknowledgements The authors thank the support to this investigation by the Brazilian agencies: CNPq, CAPES and FAPERJ and the technician Renan da Silva Guimarães.

References

1. ABIROCHAS (2018) The Brazilian sector of ornamental stones. http://abirochas.com.br/wpcontent/uploads/2018/06/abinoticias/Setor_de_Rochas_Ornamentais.pdf. Accessed 6 Aug 2018
2. Çelik MY, Sabah E (2008) Geological and technical characterization of Iscehisar (Afyon-Turkey) marble deposits and the impact of marble waste on environmental pollution. J Environ Manage 87:106–116

3. Medina G, Sáez del Bosque IF, Frías M, Sánchez de Rojas MI, Medina C (2018) Durability of new recycled granite quarry dust-bearing cements. Constr Build Mater 187:414–425
4. Sroka J, Rybak A, Sekula R, Sitars M (2016) An investigation into the influence of filler silanization conditions on mechanical and thermal parameters of epoxy resin-fly ash composites. J Polym Environ 24:298–308
5. Brito GF, Agrawal P, Araújo EM, Mélo TJA (2011) Biopolymers, biodegradable polymers and green polymers. Electron J Mater Process 6(2):127–139
6. Demartini TJC, Rodríguez RJS, Silva FS (2018) Physical and mechanical evaluation of artificial marble produced with dolomitic marble residue processed by diamond-plated bladed gang-saws. J Mater Res Tecnol. https://doi.org/10.1016/j.jmrt.2018.02.001
7. Ribeiro CEG, Rodriguez RJ, Carvalho EA (2017) Microstructure and mechanical properties of artificial marble. Constr Build Mater 149:149–155
8. Silva FS, Ribeiro CEG, Rodriguez RJ (2018) Physical and mechanical characterization of artificial stone with marble calcite waste and epoxy resin. Mater Res 21(1):1–6. https://doi.org/10.1590/1980-5373-MR-2016-0377
9. Woodworking network (2012) US Demand for Cast Polymers to Reach 251 MillionSq. Ft. in 2016. http://www.woodworkingnetwork.com/news/woodworking-industry-trends-press-releases/US-Demand-for-Cast-Polymers-to-Reach-251-Million-Sq-Ft-in-2016-149739755.htm. Accessed 22 Sept 2017
10. Ribeiro CEG, Rodriguez RJS (2015) Influence of compaction pressure and particle content on thermal and mechanical behavior of artificial marbles with marble waste and unsaturated polyester. Mater Res 18:283–290
11. Caesarstone. CaesarStone Quartz Surfaces. http://www.caesarstoneus.com/catalog/technical-specs.cfm. Accessed 1 July 2014
12. Brazilian Association of Technical Norms – ABNT (2016) ABNT NBR 7181: Soil - Grain size analysis (in Portuguese). ABNT, Rio de Janeiro
13. Carvalho EAS, Vilela NF, Monteiro SN, Vieira, CMF, Silva LC (2017) Novel artificial ornamental stone developed with quarry waste in epoxy composite. Mater. Research. 21(1). https://doi.org/10.1590/1980-5373-mr-2017-1104
14. Brazilian Association of Technical Norms – ABNT (1991) ABNT NBR MB 3388: Soil - Determination of minimum index void ratio of cohesionless soils - Method of test (in Portuguese). ABNT, Rio de Janeiro
15. Brazilian Association of Technical Norms – ABNT (2015) ABNT NBR 15845-2: Rocks for cladding Part 2: determination of bulk density, apparent porosity and water absorption (in Portuguese). ABNT, Rio de Janeiro
16. Spanish Association of Standards and Certification (2008) UNE-EN 14617-2-08: Test methods part 2: determination of the flexural strength (in Spanish). UNE-EN, Madrid
17. Lee MY, Ko CH, Chang FC, Lo SL, Lin JD, Shan MY et al (2008) Artificial stone slab production using waste glass, stone fragments and vacuum vibratory compaction. Cement Concr Compos 30(7):583–587
18. Kirgiz MS (2015) Strength gain mechanisms of blended-cements containing marble powder and brick powder. KSCE J Civil Eng 19(1):165–172
19. Chiodi Filho C, Rodriguez H (2009) Guide application of stone coverings. ABIROCHAS, São Paulo
20. Borsellino C, Calabrese L, Di Bella G (2009) Effects of powder concentration and type of resin on the performance of marble composite structures. Constr Build Mater 23(5):1915–1921

Part II
Natural Fiber Composites

Natural Fibers Reinforced Polymer Composites Applied in Ballistic Multilayered Armor for Personal Protection—An Overview

Sergio Neves Monteiro, Jaroslaw Wieslaw Drelich, Henry Alonso Colorado Lopera, Lucio Fabio Cassiano Nascimento, Fernanda Santos da Luz, Luís Carlos da Silva, Jheison Lopes dos Santos, Fábio da Costa Garcia Filho, Foluke Salgado de Assis, Édio Pereira Lima, Artur Camposo Pereira, Noan Tonini Simonassi, Michelle Souza Oliveira, Luana Cristyne da Cruz Demosthenes, Ulisses Oliveira Costa, Raphael Henrique Morais Reis and Wendell Bruno Almeida Bezerra

Abstract Natural fibers have been investigated as composites reinforcement for technological applications. Several natural fiber composites were found to present similar ballistic performance as synthetic materials, such as Kevlar™ in multilayered systems (MASs) against high-impact energy projectiles. MASs with a front ceramic followed by natural fibers, natural fabrics or natural non-woven mat such as curaua, jute, sisal, ramie, coir, piassava, fique, and sugarcane bagasse reinforcing polymer matrix composites attend the standards for personal ballistic protection against the threat of 7.62 mm ammunition. These natural fibers or fabrics or non-woven mats composites, as second MAS layer, succeed to capture the cloud of fragments resulting from the projectile impact on the ceramic layer with comparable ballistic efficiency as conventional synthetic aramid fiber composites with same thickness. In this work, the development of research dedicated to these natural fiber-based composites in MASs are overviewed and their advantages over synthetic materials are emphasized, especially the much lower cost.

Keywords Natural fibers · Polymer composites · Ballistic behaviors · Multilayered armor

---

S. N. Monteiro (✉) · L. F. C. Nascimento · F. S. da Luz · L. C. da Silva · J. L. dos Santos · F. da Costa Garcia Filho · F. S. de Assis · É. P. Lima · A. C. Pereira · N. T. Simonassi · M. S. Oliveira · L. C. da Cruz Demosthenes · U. O. Costa · R. H. M. Reis · W. B. A. Bezerra
Military Institute of Engineering, Rio de Janeiro, Brazil
e-mail: snevesmonteiro@gmail.com

J. W. Drelich
Michigan Technological University, Houghton, MI, USA

H. A. C. Lopera
University of Antioquia, Medellin, Colombia

S. Ikhmayies et al. (eds.), *Green Materials Engineering*, The Minerals, Metals & Materials Series, https://doi.org/10.1007/978-3-030-10383-5_4

Introduction

In past few decades, a marked interest has emerged for replacement of synthetic-based materials by the so-called green materials. Sustainable reasons associated with renewable and biodegradable conditions of green materials make them environmentally friendly. In this respect, natural lignocellulosic fibers, extracted from plants, are attracting attention not only because of their sustainable advantages but also due to superior technological performance as reinforcement of polymer composites [1–10]. A growing number of research works has been dedicated to investigate the properties and applications of composites reinforced with natural fibers and related fabrics. Indeed, according to international metrics disclosed by ISI Knowledge [9], the number of articles with keyword "natural fiber composites" raised from only 3 in 1990 to a projected more than 1,000 in 2017. Actually, these composites are rapidly replacing traditional synthetic fiber reinforced composites, like fiberglass, in many technological areas, mainly civil construction and automobile industries [11–13]. In particular, a rather surprising application that has been investigated in recent years is the successful use of natural fiber reinforced polymer composites as part of ballistic armor for personal protection. This is the subject of the present short overview.

Armors for Personal Ballistic Protection

A Brief Historical Background

Pre-historical armors for personal protection were primarily made of leather, wood, and strong natural fibers, such as bamboo. Figure 1 illustrates a simulated fight between eastern warriors with manual shields and split bamboo vests.

The use of metallic arms, like spears and spades, demanded personal protection with metallic armors, as the complete armor suit shown in Fig. 1b used by knights in the middle ages. Metallic armor protection was improved with the advent of firearms. Indeed, until World War II an efficient personal ballistic protection would require steel helmet and partial vest as shown in Fig. 2a. However, the increasing power of firearms, associated with higher velocity-impact-penetration (VIP) ammunition, was requiring even increasing steel thickness to guarantee body protection. This raised the weight of the armor and, consequently, reduced the wearer mobility [14]. Lighter non-metallic armors based on strong synthetic fibers such as glass, carbon, nylon, aramid, and ultra-high molecular weight polyethylene (UHMWPE) began, from the middle of past century, to be used for body protection [15]. In addition, ceramics that are less dense than metals are also being considered in modern armors for body protection [16–18]. In particular, armor vests such as the ones illustrated in Fig. 2b are today most commonly using an aramid fabric ply composite, under commercial names of Kevlar™ and Twaron™, or pressed layers of UHMWPE, under commercial names of Dyneema™ and Spectra™.

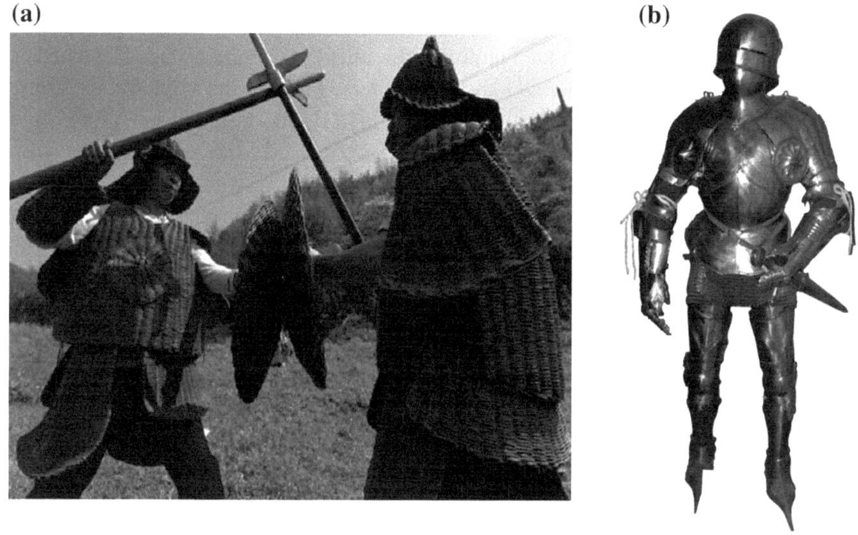

Fig. 1 **a** Fight between ancient warriors protected with natural materials for shield and vest and **b** complete steel armor suit used by knights in the middle ages

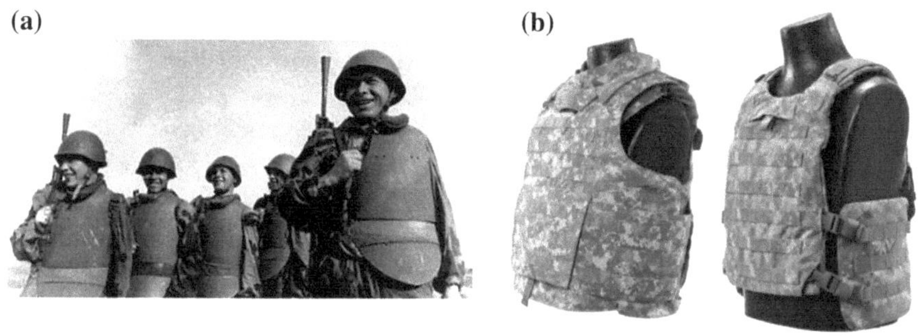

Fig. 2 Body armors **a** steel vests used by Russian soldier in WWII and **b** synthetic fiber/fabric modern vests

Natural Fibers Composites in Ballistic Armors

In recent years, research works have been investigating the possibility of applying natural fibers in ballistic armors. To the knowledge of the authors of this overview, the first publication on this subject was the 2007 work of Wambua et al. [19] on hybrid natural fibers composite and steel sheets. The authors found that the polypropylene matrix composites reinforced with 46 vol.% of either flax, hemp or jute fabrics backed by steel sheets displayed a ballistic performance superior to a plain steel armor. Other investigations on ballistic performance of hybrid natural fiber composite followed until 2013. Ali et al. [20] fabricated a hybrid laminate polyester matrix composite reinforced with both synthetic aramid fabric and natural ramie fiber for body ballistic armor. The results indicated that a laminate thickness of 25 mm is capable to protect against a projectile with impact velocity of 837 m/s, which is the same for a 7.62 mm rifle ammunition. In the work of Abidin et al. [21], the ballistic performance of a hybrid structure consisting of steel sheets intercalated with polyurethane foam matrix composite reinforced with 20 wt% of kenaf fiber was investigated. Depth of indentation in clay witness backing the 45 mm thick target indicated that a kenaf fiber polyurethane foam composite, placed between steel sheets, is able to resist the impact of high velocity 5.56 mm rifle ammunition.

Despite this aforementioned interesting ballistic performance of natural fibers and fabrics polymer composites in hybrid targets together with steel or synthetic material, up to 2015, there were doubts that a natural fiber composite could directly compete against the much stronger synthetic materials, such as Kevlar™ and Dyneema™, worldwide used in body ballistic protection as illustrated in Fig. 2b. However, in 2015 Monteiro et al. [22] revealed, for the first time, that in a multilayered armor with a front alumina-based ceramic followed by a Kevlar™ plate, the mechanism of energy absorption by this second layer depends mainly on the Kevlar's aramid fibers capacity to collect fragments resulting from the ceramic shatter after the bullet impact. In other words, the ballistic performance of a material placed after the front ceramic in a multilayered armor system (MAS) is directly associated with the effective collection of fragments and not the fiber strength. This surprising result opened the possibility of natural fibers, even being much weaker than synthetic fibers, to display comparable ballistic performance to synthetic fibers as MAS second layer. Figure 3, reproduced from [22], illustrates the collection of shattered front ceramic fragments by aramid fibers from Kevlar™ as second MAS layer.

Front Ceramic/Natural Fiber Composite Multilayered Armor

Motivation

The original results for the role played by a front ceramic MAS second layer [22] motivated a joint line of investigation between the Military Institute of Engineering (IME), Brazil, and the Michigan Technological University (Michigan Tech), USA, on second layers of natural fibers and fabric composites. From 2015 to 2018, a several articles [23–49] were published on this subject by the IME/Michigan Tech research group.

Currently, another group from the University of Antioquia (UdeA), Colombia, has joined in this research on natural fibers and fabrics composites applied in MAS for ballistic body protection. Therefore, 20 researchers including 5 seniors, 3 postdoctoral, 4 doctor graduate students, 5 master graduate students, and 3 undergraduate students are directly in this line of investigation.

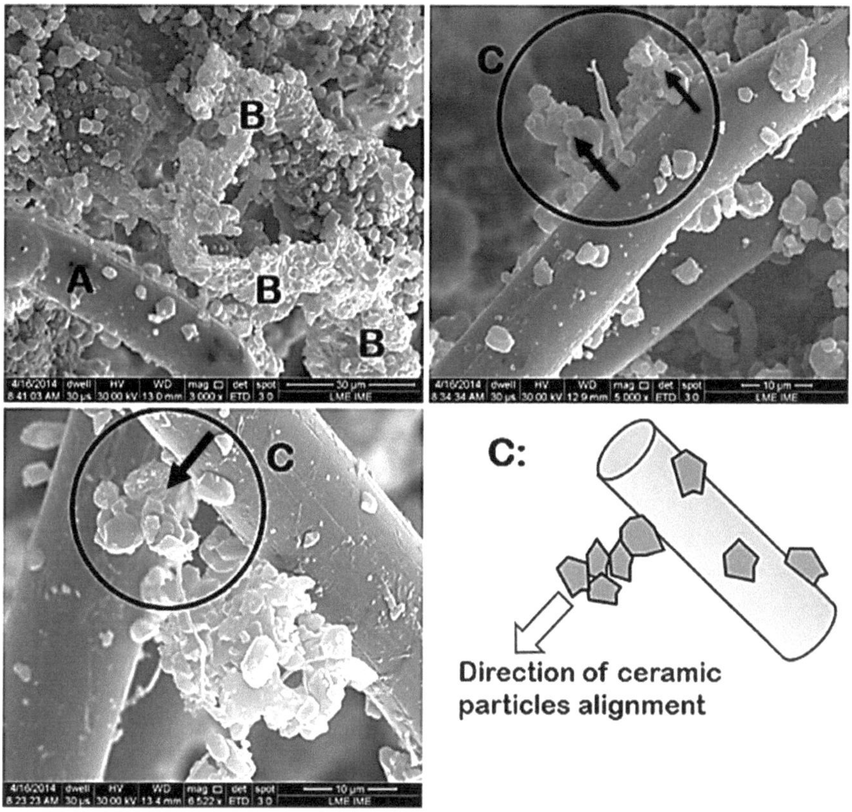

Fig. 3 SEM micrographs of the aramid fabric from tests with the multilayer armor system and schematic of alignment of ceramic fragment aggregates such as in areas circled on micrographs. A-Fiber with ~15 pct surface area covered with ceramic fragments; B-fiber covered nearly 100 pct with ceramic fragments; and C-aggregates of ceramic fragments and their alignment. Reproduced with permission [22]

Experimental

The experimental methodology regarding materials and testing procedures are specifically described in each one of the published articles on natural fibers and fabrics composites applied in MAS for ballistic body protection [23–49]. It goes beyond the scope of this overview to repeat case by case. As a summary, most investigated simple natural lignocellulosic fibers as well as natural fabrics and non-woven mats manufactured from natural fibers, were purchased from regular supplier in Brazil or abroad. Figure 4 illustrates as-received bundles of some natural fibers as well as samples of textiles and non-woven mat used so far [25, 28, 31, 36, 43, 47, 48].

Moreover, fibers were also manually extracted from supplied culm of bamboo [46] and pineapple leaf [23] as illustrated in Fig. 5a, b. In addition, waste parts of plants

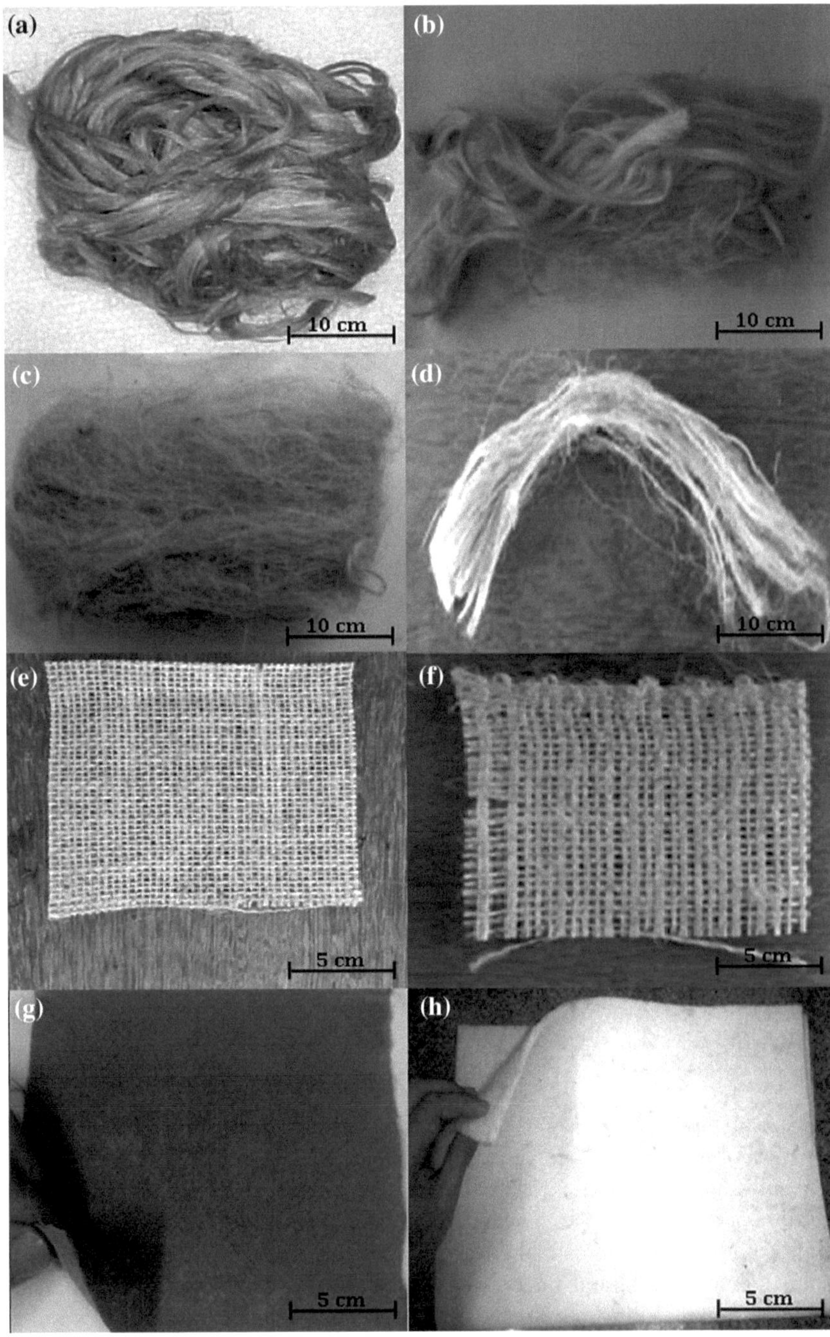

Fig. 4 Examples of **a** bundle of curaua fibers [48]; **b** bundle of sisal fibers [47]; **c** bundle of jute fibers [31]; **d** bundle of fique fiber [28]; **e** jute fabric [25]; **f** fique fabric [28]; **g** ramie fabric [43]; and **h** curaua non-woven mat [36]

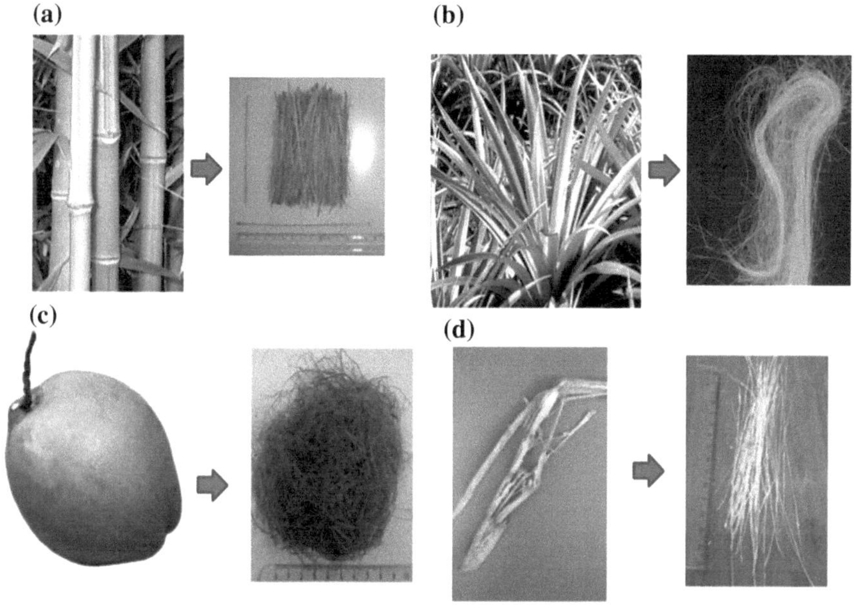

Fig. 5 Examples of fibers extracted from **a** bamboo culm [46]; **b** pineapple leaf [23]; **c** coconut shell [23]; and **d** sugarcane bagasse [44]

such as coconut shell, Fig. 5c [23] and sugarcane bagasse, Fig. 5d [44] provided natural fibers for our research work.

Composites with aforementioned fibers, fabrics, or non-woven mats were prepared by mixing different volume fractions with either epoxy or polyester thermosetting polymers as matrices. Plates with 150 × 120 × 10 mm^3 were fabricated in steel molds under pressure of 3–5 MPa.

Hexagonal-shaped tiles with side dimension of 50 mm and thickness of 10 mm were fabricated by sintering alumina (Al_2O_3) with 4 wt% of niobia (Nb_2O_5). Sheets of 5 mm thick aluminum alloy were provided to complete a multilayered armor system schematically shown in Fig. 6. The clay witness block shown in this figure has the same density and consistency of a human body to simulate the trauma after an impact against the front ceramic.

All ballistic tests were conducted at the Brazilian Army Evaluation Center (CAEx) in Rio de Janeiro using MASs with different aforementioned composites as targets against class III 7.62 × 51 mm ammunition [50].

Figure 7a shows schematically the setup used for ballistic tests together with actual MAS targets before, Fig. 7b, and after, Fig. 7c, the projectile impact. In this figure, it is also shown the "backface signature," i.e., the depth of indentation trauma caused in the clay witness. The measurement of this indentation, Fig. 7d, with a laser sensor should not exceed 44 mm, considered as a lower limit for lethal trauma [50].

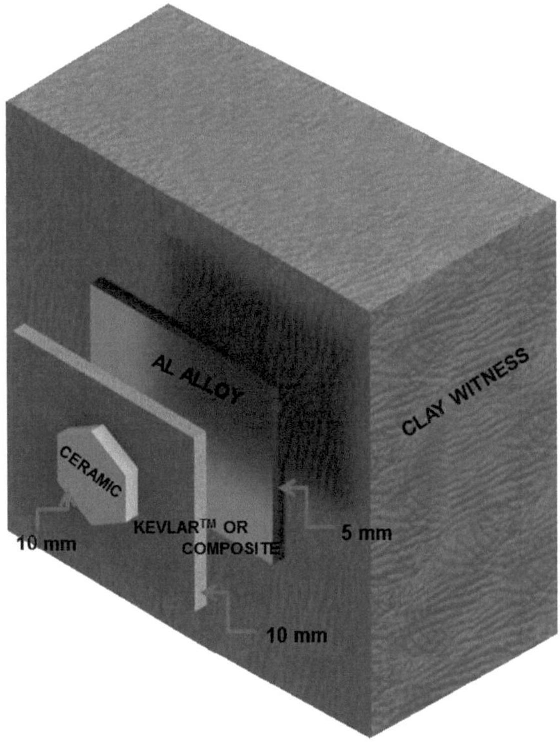

Fig. 6 Schematic illustration of a MAS with front ceramic tile followed

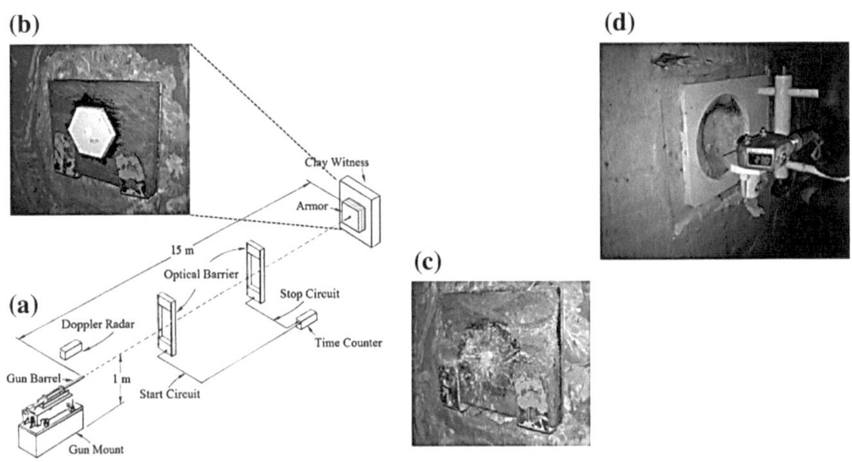

Fig. 7 **a** Schematic of the CAEx ballistic test [35]; **b** actual MAS target before [35]; **c** target after impact [35]; and **d** depth of indentation measurement with laser sensor [31]

It is worth mentioning in Fig. 7c that after all ballistic tests the hexagonal front ceramic was completely shattered resulting in a cloud of fragments. The kinetic energy from this cloud has to be partially dissipated by elsewhere discussed mechanisms [22] in the MAS second layer. The microstructure characteristic of the fragments and related capture mechanisms as well as damages produced in the MAS target by the ballistic impact were analyzed by scanning electron microscopy (SEM) coupled with energy dispersive spectroscopy (EDS) facility.

Revealed Results and Corresponding Discussion

Certainly the most important result, common to all works on ballistic performance of natural fibers or fabrics or non-woven mats reinforced polymer composites as MAS second layer [23–49], was to attend the standard requirement of indentation depth lower than 44 mm. Table 1 presents a summary of these results already published in research works performed at IME in collaboration with Michigan Tech and UdeA. In this table, the readers should note the comparison of the average indentation depth, in more than 20 ballistic tests, using the same 10 mm thickness Kevlar^{TM} as MAS second layer. It is interesting to notice that, in general, the depth of indentation found for the natural fibers/fabrics/non-woven mats composites as MAS second layers are comparable to that of MAS with Kevlar^{TM}. In addition to these articles published by the research partnership between IME, Michigan Tech and UdeA, it is important to mention a recent review by Benzait and Trabzon [51] in which a significant part is dedicated to our articles [29, 35, 37, 40–44, 46–49].

Another relevant point regarding the natural fibers/fabrics/non-woven mats reinforced polymer composites as MAS second layer in ballistic tests is their shock wave impedances. Indeed, owing to the comparatively lower density of these composites as compared to the front ceramic, the compressive shock wave from the projectile impact is reflected as a tensile wave at the interface [47]. This contributes to shatter the front ceramic and to increase the amount of dissipated energy. Moreover, since these composites have lower density than Kevlar^{TM}, their impedance is also lower and corresponding higher-amplitude tensile wave is reflected. In principle, composites such as those shown in Table 1 might absorb even greater ballistic impact energy than Kevlar^{TM}. This is apparently the case of 30 vol.% jute fabric/polyester and 30 vol.% ramie fabric/epoxy composites in Table 1. For specific calculation of shock wave impedance, the reader is referred to previous works [24, 25, 35, 44].

A final point worth discussing in this overview is the mechanism of ballistic energy dissipation by natural fibers/fabrics/non-woven mats composites as MAS second layers. In all investigated composites, evidence of fragments capture, similar to the mechanisms proposed for Kevlar^{TM} as MAS second layer in Fig. 3 [22], was found for both the natural fiber (also in fabric or mat) and the polymer matrix. Figure 8 illustrates the capture of ceramic and projectile fragments, resulting from the ballistic impact , by the polyester matrix, Fig. 8a [27]; the epoxy matrix, Fig. 8b [35]; curaua

Table 1 Depth of indentation of MAS with natural fibers/fabrics/non-woven mats composites and same thickness Kevlar™ for comparison

MAS second layer	Depth of indentation (mm)	Reference
Kevlar™	23 ± 3	[49]
30 vol.% Curaua fiber/epoxy composite	22 ± 3	[48]
30 vol.% curaua non-woven fiber/epoxy composite	28 ± 3	[36]
30 vol.% Curaua fiber/polyester composite	22 ± 2	[35]
30 vol.% pineapple leaf fibers/epoxy composite	18.2 ± 2.7	[23]
30 vol.% jute fabric/polyester composite	17 ± 2	[25]
30 vol.% jute fabric/epoxy composite	21 ± 3	[49]
30 vol.% jute non-woven mat/polyester composite	24 ± 7	[31]
30 vol.% coir fiber/epoxy composite	31.6 ± 2.7	[23]
30% coir non-woven mat/epoxy composite	24 ± 6	[29]
30 vol.% bagasse fiber/epoxy composite	21 ± 1	[44]
30 vol.% ramie fibers/epoxy composite	18 ± 2	[27]
30 vol.% ramie fabric/epoxy composite	17 ± 1	[43]
20 vol.% fique fabric/polyester composite	15 ± 3	[28]
30 vol.% sisal fiber/polyester composite	23	[24]
30 vol.% sisal fiber/epoxy composite	18 ± 2	[47]
30 vol.% giant bamboo fiber/epoxy composite	18 ± 2	[46]
30 vol.% mallow fiber/epoxy composite	22.4 ± 1.3	[33]
30 vol.% malow fabric/epoxy composite	21.48 ± 1.64	[37]
30 vol.% hybrid mallow/jute fabric (70/30)/epoxy composite	23.16 ± 3.12	[37]
30 vol.% hybrid mallow/jute fabric (50/50)/epoxy composite	23.7 ± 2.4	[42]
30 vol.% raw bagasse/epoxy composite	39 ± 8	[44]

fibers, Fig. 8c [35]; ramie fabric, Fig. 8d [43]; sugarcane bagasse fiber, Fig. 8e [44]; and jute non-woven mat, Fig. 8f [31]. Based on the similar capture mechanisms by the aramid fibers [22], it is also proposed that both the polymer matrix and the nature fiber, in whatever configuration, captures fragments from front ceramic projectile impact by incrustation, Van der Waals forces and short-living surface static charges generated during fiber-ceramic interaction.

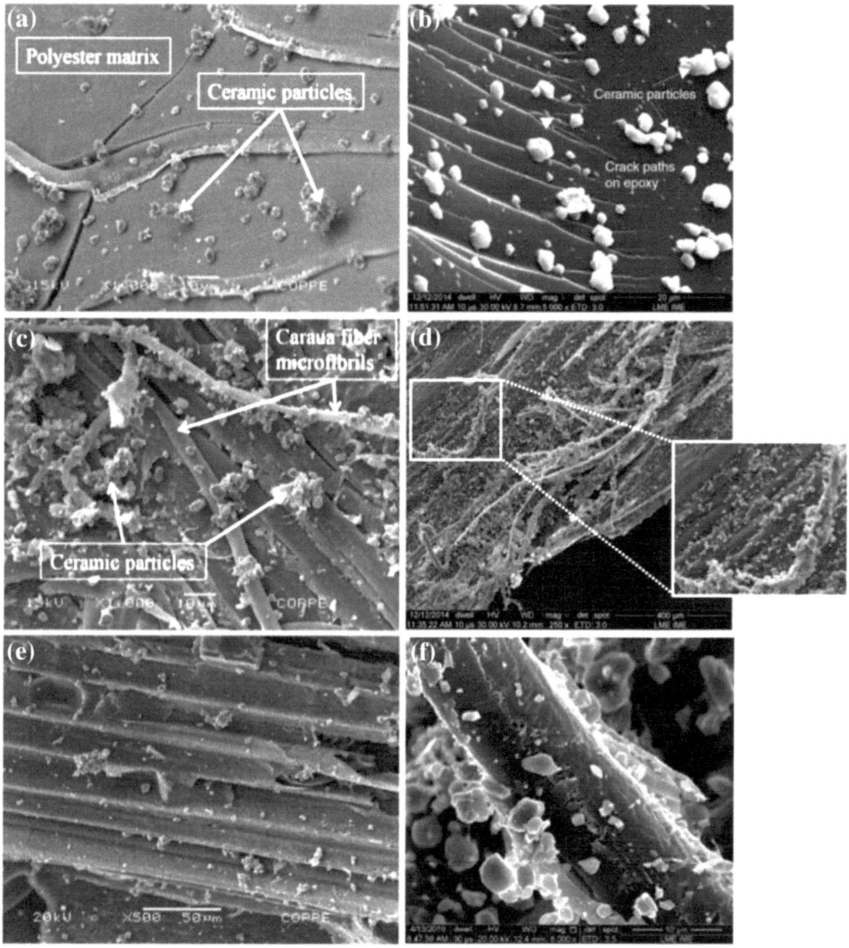

Fig. 8 Capture of ceramic/projectile fragments by **a** polyester matrix [27]; **b** epoxy matrix in [35]; **c** curaua fibers in [35]; **d** ramie fabric in [43]; **e** sugarcane bagasse fiber in [44]; and **f** jute fibers in non-woven mat in [31]

Summary and Conclusions

A brief overview shows that research works in combined partnership between the Military Institute of Engineering (IME), in Brazil, the Michigan Technological University (Michigan Tech), in USA, and the University of Antioquia (UdeA), in Colombia, revealed for the first time that natural fiber reinforced polymer composites have similar ballistic performance of commonly worldwide used synthetic materials.

NIJ standard ballistic tests were conducted at the Brazilian Army Assessment Center (CAEx), using composites reinforced with either natural fibers, or natural fabrics, or non-woven mat as second layer following front alumina-based ceramic multilayered armor system (MAS). These tests disclosed indentation depths comparable to that found for same 10 mm thick Kevlar™ as MAS second layer.

This surprising natural fiber/fabric/non-woven mat composite ballistic performance in comparison with the much stronger synthetic aramid fibers in the Kevlar™, was explained by the Michigan Tech researcher as a consequence of fragment capture mechanisms.

All investigated epoxy or polyester composites reinforced with different natural fibers/fabrics/non-woven mats, including fique fibers and fabrics from UdeA, displayed not only similar ballistic performance as MAS second layer but also same capture mechanisms. These were proposed to be comparable to those for aramid fibers involving fragment particles incrustation, Van der Waals forces, and short-living surface static changes generated during the natural fiber interaction with fragments after the ballistic impact.

Acknowledgements The authors thank the support of this investigation by the Brazilian agencies: CNPq, FAPERJ, and CAPES.

References

1. Mohanty AK, Misra M, Drzal LT (2002) Sustainable bio-composites from renewable resources: opportunities and challenges in the green materials world. J Polym Environ 10(1):19–26
2. Netravali AN, Chabba S (2003) Composites get greener. Mater Today 6(4):22–29
3. Crocker J (2008) Natural materials innovative natural composites. Mater Technol 23(3):174–178
4. John MJ, Thomas S (2008) Biofibers and biocomposites. Carbohydr Polym 71(3):343–364
5. Monteiro SN, Lopes FPD, Ferreira AS, Nascimento DCO (2009) Natural fiber polymer matrix composites: cheaper, tougher and environmentally friendly. JOM 61:17–22
6. Monteiro SN, Lopes FPD, Barbosa APB, Bevitori AB, Da Silva ILA, Da Costa LL (2011) Natural lignocellulosic fibers as engineering materials—an overview. Metall Mater Trans A 42A:2963–2974
7. Faruk O, Bledzki AK, Fink HP, Sain M (2012) Biocomposites reinforced with natural fibers. Prog Polym Sci 37(11):1555–1596
8. Thakur VK, Thakur MK, Gupta RK (2014) Review: raw natural fibers based polymer composites. Int J Polym Anal Charact 19(3):256–271
9. Guven O, Monteiro SN, Moura EAB, Drelich W (2016) Re-emerging field of lignocellulosic fiber-polymer composites and ionizing radiation technology in then formulation. Polym Rev 56:702–736
10. Pickering KL, Effendy MGA, Le TM (2016) A review of recent developments in natural fibre composites and their mechanical performance. Compos Part A: Appl Sci Manuf 86:98–112
11. Holbery J, Houston D (2006) Natural fiber reinforced polymer composites in automotive applications. JOM 58(11):80–86
12. Zah R, Hischier R, Leal AL, Brown I (2007) Curaua fibers in automobile industry: a sustainable assessment. J Clean Prod 15(11–12):1032–1340

13. Thomas N, Paul AS, Pothan LA, Deepa B (2011) Natural fibers: structure, properties and application. In: Kalia S, Kath BS, Kawr I (eds) Cellulose fibers: bio- and nano-polymer composites, 1st edn. Springer, Berlin/Heidelberg, Germany, pp 3–42
14. Büger D, Faria AR, Almeida SFM, Melo FCL, Donadon MV (2012) Ballistic impact simulation of an armour-piercing projectile on hybrid ceramic/fiber reinforced composite armours. Int J Impact Eng 43:63–77
15. Wang L, Kanesalingam S, Nayak R, Padhye R (2014) Recent trends in ballistic protection. Textiles and Light Ind Sci Technol 3:37–47
16. Medvedovski E (2006) Lightweight ceramic composite armour system. Adv Appl Ceram 105(5):241–245
17. Medvedovski E (2010) Ballistic performance of armour ceramics: influence of design and structure. Part 1. Ceram Int 36(7):2103–2115
18. Serjouei A, Chi R, Zhang Z, Sridhar I (2015) Experimental validation of BLV model on bi-layer ceramic-metal armor. Int J Impact Eng 77:30–41
19. Wambua P, Vangrimde B, Lomov S, Verpoest I (2007) The response of natural fibre composites to ballistic impact by fragment simulating projectiles. Compos Struct 77(2):232–240
20. Ali A, Shaker ZR, Khalina A, Sapuan SM (2011) Development of anti-ballistic board from ramie fiber. Polymer-Plastics Technol Eng 50(6):622–634
21. Abidin MHZ, Mohamad MAH, Zaidi AMA, Mat WAW (2013) Experimental study on ballistic resistance of sandwich panel protection structure with kenaf foam as a core material against small arm bullet. Appl Mech Mater 315:612–615
22. Monteiro SN, Lima EP Jr, Louro LHL, Silva LC, Drelich JW (2014) Unlocking function of aramid fibers in multilayered ballistic armor. Metall Mat Trans A 46(1):37–40
23. Luz FS, Ramos FJHTV, Nascimento LFC, Figueiredo AB-HS, Monteiro SN (2018) Critical length and interfacial strength of PALF and coir fiber incorporated in epoxy resin matrix. J Mater Res Technol. https://doi.org/10.1016/j.jmrt.2018.04.025
24. Braga FO, Bolzan LT, Ramos FJHTV, Monteiro SN, Lima EP Jr, Silva LC (2018) Ballistic efficiency of multilayered armor systems with sisal fiber polyester composites. Mater Res 20(suppl 2):767–774
25. Monteiro SN, Pereira AC, Ferreira C, Lima EP Jr (2018) Performance of plain woven jute fabric-reinforced polyester matrix composite in multilayered ballistic system. Polymers 10(3):230–239
26. Silva AO, Monsores KGC, Oliveira SSA, Weber RP, Monteiro SN (2018) Ballistic behavior of a hybrid composite reinforced with curaua and aramid fabric subjected to ultraviolet radiation. J Mater Res Technol (accepted for publication)
27. Braga FO, Milanezi TL, Monteiro SN, Louro LHL, Gomes AV, Lima Jr EP (2018) Ballistic comparison between epoxy-ramie and epoxy-aramid composites in Multilayered Armor Systems. J Mater Res Technol. https://doi.org/10.1016/j.jmrt.2018.06.018
28. Monteiro SN, Assis FS, Ferreira CL, Simonassi NT, Weber RP, Oliveira MS, Lopera HAC, Pereira AC (2018) Fique Fabric: a promising reinforcement for polymer composites. Polymers 10(3):246–255
29. Luz FS, Monteiro SN, Lima ES, Lima EP Jr (2018) Ballistic application of coir fiber reinforced epoxy composite in multilayered armor. Mater Res 20(Suppl. 2):23–28
30. Braga FO, Lopes PHLM, Luz FS, Lima Jr EP, Monteiro SN (2018) Influence of the areal density of layers in the ballistic response of a multilayered armor system using box-behnken statistical design. In: Li B et al. (eds) Characterization of minerals, metals, and materials 2018. The Minerals, Metals & Materials Series. Cham, Switzerland: Springer Int Publishing 557–564
31. Assis FS, Pereira AC, Garcia Filho FC, Lima Jr EP, Monteiro SN, Weber RP (2018) Performance of jute non-woven mat reinforced polyester matrix composite in multilayered armor. J Mater Res Technol. https://doi.org/10.1016/j.jmrt.2018.05.026

32. Nascimento LFC, Louro LHL, Monteiro SN, Braga FO, Luz FS, Santos JL, Marçal RLSB, Freitas HCO, Lima Jr EP (2018) Limit speed analysis and absorbed energy in multilayer armor with epoxy composite reinforced with mallow fibers and mallow and jute hybrid fabric. In: Li B et al (eds) The minerals, metals & materials series. Cham, Switzerland: Springer International Publishing 597–604
33. Nascimento LFC, Louro LHL, Monteiro SN, Lima EP Jr, Luz FS (2017) Mallow fiber-reinforced epoxy composites in multilayered armor for personal ballistic protection. JOM 69:2052–2056
34. Nascimento LFC, Holanda LIF, Louro LHL, Monteiro SN, Gomes AV, Lima EP Jr (2017) Natural mallow fiber-reinforced epoxy composite for ballistic Armor against class III-A ammunition. Metall Mater Trans A 48:4425–4431
35. Monteiro SN, Braga FO, Lima EP Jr, Louro LHL, Drelich JW (2017) Promising curaua fiber-reinforced polyester composite for high-impact ballistic multilayered armor. Polymer Eng Sci 57:947–954
36. Braga FO, Cabral AC, Lima Jr EP, Monteiro SN, Assis FS (2017) Curaua non-woven fabric composite for ceramic multilayered Armors: a lightweight, natural, and low cost alternative for Kevlar^{TM}. In: Meyers MA et al (eds) Pan american materials conference. The Minerals, Metals & Materials Series. Springer International Publishing 1:339–346
37. Nascimento LFC, Louro LHL, Monteiro SN, Gomes AV, Lima Jr EP, Marçal RLSB (2017) Ballistic performance in multilayer armor with epoxy composite reinforced with malva fibers. In: Meyers MA et al (eds) Pan American materials conference. the minerals, metals & materials series. Springer International Publishing 1:331–338
38. Nascimento LFC, Holanda LIF, Louro LHL, Monteiro SN, Gomes AV, Lima Jr EP (2017) Evaluation of ballistic armor behavior with epoxy composite reinforced with malva fibers. In: Li B et al (eds) Characterization of minerals, metals, and materials 2018. The Minerals, Metals & Materials Series. Cham, Switzerland: Springer Int Publishing 1:647–655
39. Monteiro SN, Figueiredo ABS, Lima ES, Weber RP, Louro LHL, Matos MAJ, Nascimento LFC, Braga FO (2017) The role of sintered Al_2O_3-Nb_2O_5 front plate on the ballistic performance of multilayered armors. Mater Sci Forum (online) 899:329–334
40. Braga FO, Bolzan LT, Lima EP Jr, Monteiro SN (2017) Performance of natural curaua fiber-reinforced polyester composites under 7.62 mm bullet impact as a stand-alone ballistic armor. J Mater Res Technol 6:323–328
41. Braga FO, Bolzan LT, Luz FS, Lopes PHLM, Lima EP Jr, Monteiro SN (2017) High energy ballistic and fracture comparison between multilayered armor systems using non-woven curaua fabric composites and aramid laminates. J Mater Res Technol 6:417–422
42. Nascimento LFC, Louro LHL, Monteiro SN, Gomes AV, Marçal RLSB, Lima EP Jr, Margem JI (2017) Ballistic performance of mallow and jute natural fabrics reinforced epoxy composites in multilayered armor. Mater Research-Ibero-Am J Mater 1:399–403
43. Monteiro SN, Milanezi TL, Louro LHL, Lima EP Jr, Braga FO, Gomes AV, Drelich JW (2016) Novel ballistic ramie fabric composite competing with Kevlar- fabric in multilayered armor. Mater Des 96:263–269
44. Monteiro SN, Candido VS, Braga FO, Bolzan LT, Weber RP, Drelich JW (2016) Sugarcane bagasse waste in composites for multilayered armor. Eur Polymer J 78:173–185
45. Garcia Filho FC, Monteiro SN (2018) Piassava Fiber as na epoxy matrix composite reinforcement for ballistic armor applications. JOM (Accepted for publication). https://doi.org/10.1007/s11837-018-3148-x
46. Cruz RB, Lima EP Jr, Monteiro SN, Louro LHL (2015) Giant Bamboo fiber reinforced epoxy composite in multilayered ballistic armor. Mater Res 18:70–75
47. Rohen LA, Margem FM, Monteiro SN, Vieira CMF, Araujo BM, Lima ES (2015) Ballistic efficiency of an individual epoxy composite reinforced with sisal fibers in multilayered armor. Mater Res 18:55–62
48. Monteiro SN, Louro LHL, Trindade W, Elias CN, Ferreira CL, Lima ES et al (2015) Natural curaua fiber-reinforced composites in multilayered ballistic armor. Metall Mater Trans A 46(10):4567–4577

49. Luz FS, Lima EP Jr, Louro LHL, Monteiro SN (2015) Ballistic test of multilayered armor with intermediate epoxy composite reinforced with jute fabric. Mater Res 18:170–177
50. National Institute of Justice NIJ Standards 0101.04 (2000) Ballistic Resistance of Body Armor. US Dept of Justice - Office of Justice Programs
51. Benzait Z, Trabzon L (2018) A review of recent research on materials used in polymer-matrix composites for body armor application. J Compos Mater 52(23):3241–3263

Structure–Property Relation of Epoxy Resin with Fique Fibers: Dynamic Behavior Using Split-Hopkinson Pressure Bar and Charpy Tests

Julian Rua, Mario F. Buchely, Sergio Neves Monteiro and Henry A. Colorado

Abstract The main objective of this work is to study polymer matrix composites from epoxy resin reinforced with fique natural fibers. Laminate composite samples with 0.0, 4.0 and 8.0 wt% of fibers in a fabric configuration were fabricated. No further treatment was used in order to make the process as simple and inexpensive as possible for diverse structural applications. Microstructural characterization was analyzed by scanning electron microscopy. Impact tests evaluated by Charpy and Split-Hopkinson pressure bar (SHPB) tests were also conducted.

Keywords Fique fibers · Epoxy resin · Composite materials · Split-Hopkinson bar · Impact · Charpy test

Introduction

Fique is a Colombian natural fiber whose fibers are typically used for sacks, crafts and to contain slopes [1]. It has recently shown a lot of interest potential as reinforcement fiber for polymer composites in high-performance applications [2–5]. Fique is a fiber that grows in the leaves of the fique plant, Furcraea andina, a xerophytic monocot native to Andean regions of Colombia, Ecuador, and Peru. From these countries,

J. Rua · H. A. Colorado (✉)
CCComposites Laboratory, Universidad de Antioquia UdeA,
Calle 70 N°. 52-21, Medellín, Colombia
e-mail: henry.colorado@udea.edu.co

M. F. Buchely
Materials Science and Engineering, Missouri University of Science
and Technology, Rolla, MO 65409, USA

S. N. Monteiro
Military Institute of Engineering, IME, Praça General Tibúrcio 80,
Urca, Rio de Janeiro 22290-270, Brazil

H. A. Colorado
Facultad de Ingenieria, Universidad de Antioquia, Bloque 20,
Calle 67 no. 53 - 108, Medellin, Colombia

S. Ikhmayies et al. (eds.), *Green Materials Engineering*, The Minerals,
Metals & Materials Series, https://doi.org/10.1007/978-3-030-10383-5_5

fique was extended to Venezuela and Brazil. The common application in Colombia for the fique fibers is as cabuya, a bundle of fique fibers used as raw materials for the fabrication of sacks and packages for agriculture and coffee grains.

Natural fibers are commonly used as composite reinforcement in many applications that includes ballistic protection [6–9] and impact [10], composites for construction and building materials [11], automobile industry [12], arts and innovation [13, 14], and cement-based materials [15, 16]. Fique has proven to be as competitive as traditional materials for high-performance applications [2–5], including the extreme conditions seen in ballistic protection [4].

The present investigation has the objective of investigate the potential of fique fibers reinforcing epoxy resin evaluated under impact conditions giving by the Charpy and Split-Hopkinson pressure bar (SHPB) tests. Scanning electron microscopy of the fibers is also included.

Materials and Methods

Samples were fabricated with a generic commercial epoxy resin and a transparent hardener. The density of this resin is 1.103 g/cm^3 at 20 °C. The viscosity of the resin is 700 at 1500 mPas, and the gel time is between 30 and 35 min. The hardener has a density of 1.006 g/cm^3 and a higher viscosity than resin (200–400 mPas). A total of 24 wt% of hardener is used with respect to the resin contents.

Fique fibers were also commercially obtained in bundle of fibers conforming fique fabric known as *cabuya* bags with fibers oriented 0°/90° and 45°/−45° alternately. For the elaboration of the composite, ten pieces of *cabuya* fabric were placed in a mold with the same shape, and after being previously impregnated manually with the resin by using a brush; see Fig. 2. The mold was kept closed under pressure until the curing was complete. Thereafter, samples were released from molds and kept in open air for 24 h. The final sample looked as tile, from which diverse samples sizes were cut for Charpy tests by following the ASTM E23-00, with samples dimensions of 55 × 10 × 10 mm^3. The sample groove was made with a V type. Compression samples were fabricated with a plug cutter of ½″, providing samples with 10 mm in height and 12 mm in diameter.

Conventional compression of Split-Hopkinson pressure bar (SHPB) tests was carried out in epoxy resin and composite samples at Laboratories of Missouri S&T. Cylindrical samples (approximately 12.5 mm diameter and 11 mm long) were cut by metallic punches from solid epoxy resin and final composite material. SHPB system consists of 7075-T6 aluminum compression (12.7 mm diameter and 1270 mm long) bars and a striker (12.7 mm diameter and 203 mm long) bar made from the same 7075-T6 aluminum. Striker was launched using argon gas cannon at different velocities to obtain different strain rate responses. A cylindrical pulse shaper (6.3 mm diameter and 2 mm long) made from oxygen-free copper was used to filter and modulate the incident pulse. Samples were placed between compression bars using Molykote grease to reduce friction and dissipate any heat.

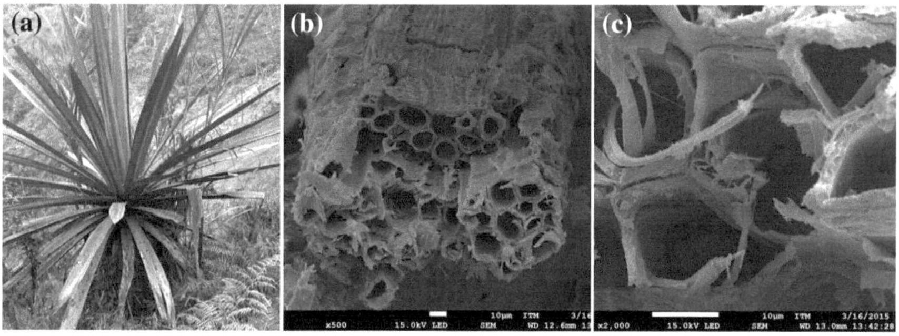

Fig. 1 Fique plant and detail of the fibers in SEM

Fig. 2 Manual modeling and test samples for Charpy and compression test

For the scanning electron microscopy (SEM) examinations, samples were sputtered in a Hummer 6.2 system (15 mA AC for 30 s) depositing a thin film of pure gold. The SEM was a JEOL JSM 6490LV used in high vacuum mode.

Results

A typical fique plan in a slope as typically used in the rural areas of Colombia is shown in Fig. 1a. The scanning electron microscopy (SEM) of the extracted fibers is shown in Fig. 1b, with a higher magnification in Fig. 1c. The fiber is quite porous, and therefore, this has a strong influence in its density, weight, and flexibility properties.

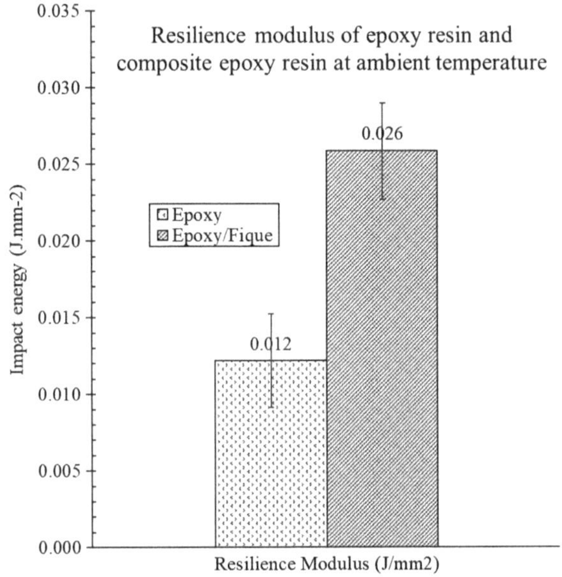

Fig. 3 Charpy test for epoxy resin and for epoxy composite at room temperature

The impact behavior in the matrix material is observable in Fig. 3, where an increased energy of absorption is added to the material using fique as reinforcement. Fique acts as a stress–strain distributor inside the sample.

SHPB Tests

Figure 4 shows an example of the recorded data from the SHPB system for the epoxy resin sample (Fig. 4a) and for the composite sample (Fig. 4b). Dynamic equilibrium was checked to ensure ideal conditions for the experiments. After that, strain rates, strain, and stresses were calculated as follows:

$$\dot{\varepsilon} = \frac{-2C_0}{l_s}\varepsilon_r,$$

$$\varepsilon = \frac{-2C_0}{l_s}\int \varepsilon_r,$$

$$\sigma = \frac{A_0}{A_s}E_0\varepsilon_t;$$

where C_0, A_0, E_0 are the sound speed and cross-sectional area and elastic modulus of the compression bars; l_s, A_s are the initial length and cross-sectional area of the specimen; ε_r is the reflected wave; and ε_t is the transmitted wave.

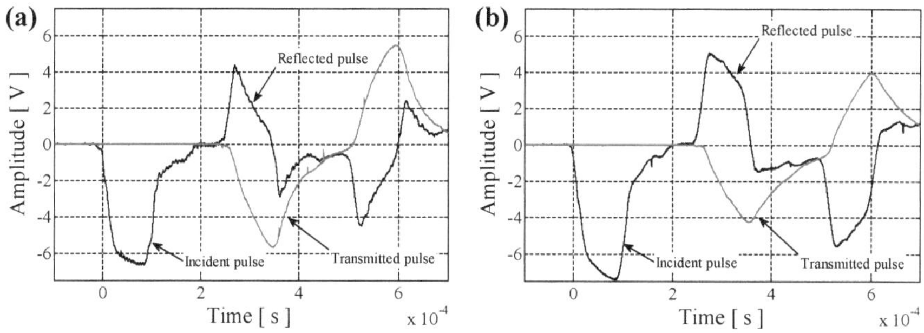

Fig. 4 Amplitude time signals recorded from the conventional SHPB tests: **a** Epoxy resin, specimen 1 and **b** Composite, specimen 1

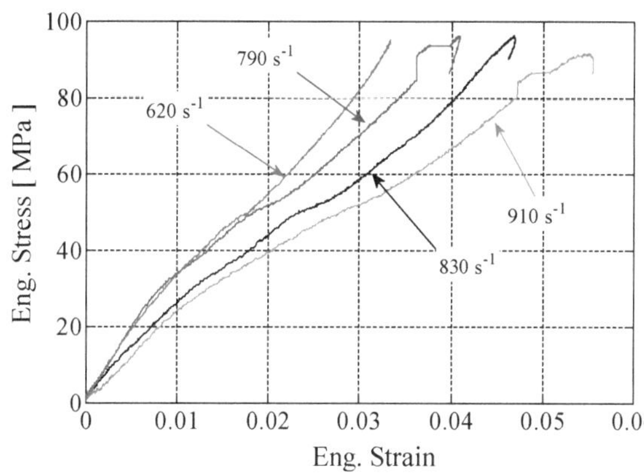

Fig. 5 Engineering stress–strain curves for epoxy resin specimens obtained by the SHPB at different strain rates. It is shown in the average strain rate during the compression test

Table 1 Summary of SHPB tests in epoxy resin specimens

Sample	Striker vel. (m/s)	Average strain rate (s^{-1})	Max. strength (MPa)	Elongation (%)
1	15.28	620	95	0
2	16.61	790	96	Failure
3	18.26	830	96	Failure
4	18.57	910	92	Failure

Figure 5 shows the stress–strain curves for the epoxy resin at different strain rates, obtained after postprocessing the SHPB data, and Fig. 6 shows the stress–strain curves for the composite material. Any numerical filtering was used for processing data and that is why some noise is present in the plots; however, this level of noise is common in SHPB systems. Most important features of stress–strain results are summarized in Tables 1 and 2. Additionally, Figs. 7 and 8 show the final shape of tested specimens after SHPB.

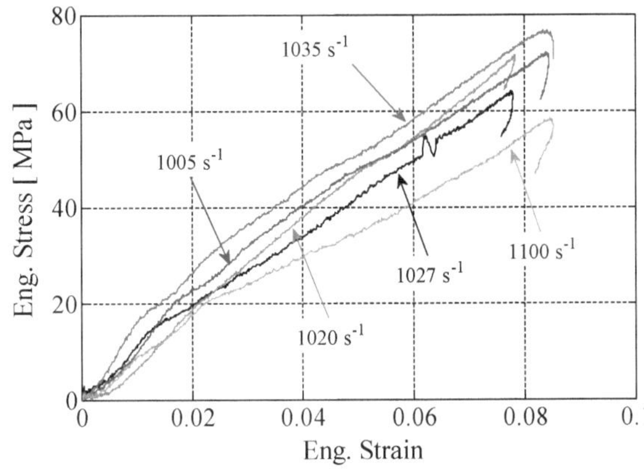

Fig. 6 Engineering stress–strain curves for composite specimens obtained by the SHPB at different strain rates. It is shown in the average strain rate during the compression test

Table 2 Summary of SHPB tests in composite specimens

Sample	Striker vel. (m/s)	Average strain rate (s^{-1})	Max. strength (MPa)	Elongation (%)
1	16.22	1100	58	0
2	18.96	1027	64	0.6
3	19.38	1020	71	0.9
4	21.09	1005	72	1.0
5	22.21	1035	76	1.5

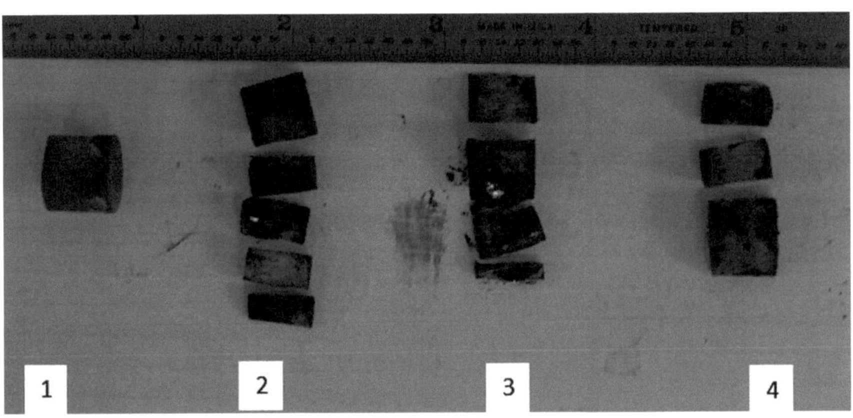

Fig. 7 Final shape of samples after SHPB tests. Scale in Inches

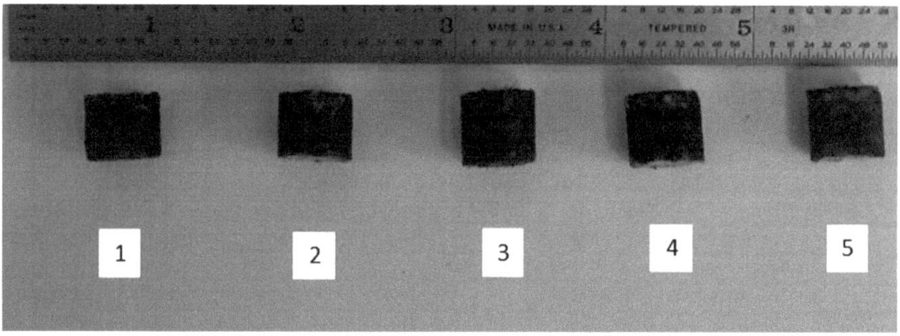

Fig. 8 Final shape of samples after SHPB tests. Scale in Inches

From these results, notice that the flow behavior is almost linear for the epoxy resin in the range of tested strain rates. Maximum strength is almost constant (approxiamately 94 MPa); however, the epoxy resin seems to lose rigidity at higher strain rates. Additionally, at striker velocities higher than 15.4 m/s, the resin fractures without any evidence of ductility (see Fig. 7 and Table 1).

Related to the composite SHPB results, it seems that the flow behavior of this material is almost bilinear, showing a yield point of approximately 20 MPa and flowing to maximum strength up to 76 MPa in some strain-rate conditions. This result shows a reduction of approxiamately 20% when compared to the matrix (epoxy resin). However, the reinforced material added some ductility to the material, up to 1.5% elongation. This elongation was not possible to obtain by only the epoxy resin matrix. Additionally, as seen in Fig. 8, composites samples were not failed, even at higher striker velocities than those tested in the matrix. There is no evident relation between the stress flow and the strain rate for the composite, and actually it seems some random behavior in the tested range, which is usually normal condition in composites. Therefore, more samples and more experiments should be done to obtain a better understanding of the relation of the strain rates and the flow stress (Table 2).

Conclusion

In this investigation, fique fibers were used as reinforcement of an epoxy resin material Charpy impact and compression Split-Hopkinson pressure bar (SHPB) revealed new limits of these materials under two different regimens. This opens up new applications that deserve the use of this composite material.

References

1. Echeverri RDE, Montoya LMF, Velásquez MRG (2015) Fique en Colombia. Instituto Tecnológico Metropolitano
2. Gañán P, Mondragon I (2003) Thermal and degradation behavior of fique fiber reinforced thermoplastic matrix composites. J Therm Anal Calorim 73(3):783–795
3. Neves Monteiro S, Salgado de Assis F, Ferreira CL, Tonini Simonassi N, Pondé Weber R, Souza Oliveira M, … & Camposo Pereira A (2018) Fique fabric: a promising reinforcement for polymer composites. Polymers 10(3):246
4. Pereira AC, Monteiro SN, Assis FS, Colorado HA (2017) Charpy toughness behavior of fique fabric reinforced polyester matrix composites. In: Characterization of minerals, metals, and materials 2017. Springer, Cham, pp. 3–9
5. Teles MCA, Altoé GR, Amoy Netto P, Colorado H, Margem FM, Monteiro SN (2015) Fique fiber tensile elastic modulus dependence with diameter using the Weibull statistical analysis. Mater Res 18:193–199
6. Pickering K (ed) (2008) Properties and performance of natural-fibre composites. Elsevier
7. Neves Monteiro S, de Assis FS, Ferreira CL, Simonassi NT, Weber RP, Oliveira MS, Colorado HA, Pereira AC (2018) Fique fabric: a promising reinforcement for polymer composites. Polymers 10(3):246
8. Pereira AC, Monteiro SN, Assis FS, Colorado HA (2017) Charpy toughness behavior of fique fabric reinforced polyester matrix composites. In: Characterization of minerals, metals, and materials 2017. Springer, Cham, pp 3–9
9. Teles, MCA, Altoé GR, Netto PA, Colorado H, Margem FM, Monteiro SN (2015) Fique fiber tensile elastic modulus dependence with diameter using the Weibull statistical analysis. Mater Res 18:193–199
10. Colorado HA, Navarro A, Prikhodko SV, Yang JM, Ghoniem N, Gupta V (2013) Ultrahigh strain-rate bending of copper nanopillars with laser-generated shock waves. J Appl Phys 114(23):233510
11. Coloradoa HA, Coloradoa SA, Buitrago-Sierrab R (2015) Portland cement with luffa fibers. In: Developments in strategic ceramic materials: a collection of papers presented at the 39th international conference on advanced ceramics and composites, January 25–30, 2015, Daytona Beach, Florida, vol. 604, p. 103. John Wiley & Sons
12. Alves C, Silva AJ, Reis LG, Freitas M, Rodrigues LB, Alves DE (2010) Ecodesign of automotive components making use of natural jute fiber composites. J Clean Prod 18(4):313–327
13. Fibras vegetales: elemento básico de las artesanías. http://www.artesaniasdecolombia.com.co/PortalAC/C_noticias/fibras-vegetales-elemento-basico-de-las-artesanias_5079. Website visited in October 2018
14. Silva C, Vélez G, Colorado HA (2017) Patina in the construction of the poetic bronze image: science of materials, art and philosophy. Herit Sci 5(1):36
15. Ribeiro MG Sá, Vasconcelos RP, Vieira RK, Vieira AK, Bittencourt E, Ribeiro RAS (2009) Building of a sustainable ecological village in the amazon-related projects and activities. Chem Eng Trans 17: 343–348
16. Quintero-Dávila M, Monteiro SN, Colorado HA (2018) Composites of Portland cement and fibers of Guadua angustifolia Kunth from Colombia. J Compos Mater 0021998318792297

Comparison of the Impact Properties of Composites Reinforced by Natural Fibers

Felipe Perissé Duarte Lopes and Carlos Mauricio Fontes Vieira

Abstract The search for new green materials is the motivation to study eco-friendly composites formulated with natural resin-based and natural fibers. The search for new green materials is the motivation to study eco-friendly composites formulated with natural resins and natural fibers, but also mixing synthetic and natural materials. This work evaluated the impact properties of polymeric composites reinforced by natural fibers, using six different types of natural fibers and two resins. Specimens were made with up to 30% vol of natural fibers and tested by impact testing. The results showed that it is possible to develop eco-friendly composites with good impact properties, using natural fibers as reinforcement in a composite. These new eco-friendly composites can be used at automobile and civil industries.

Keywords Impact properties · Natural fibers · Polymeric composites

Introduction

Polymeric composites are structural molding materials formed by a continuous polymeric phase called matrix, which is reinforced by a discontinuous phase, reinforcement, which are added physic–chemically before the polymerization (curing) process. The discontinuous phase, or reinforcement, usually consists in fibers, which can be glass, aramid, carbon, natural lignocellulosics, or animal origin, depending on the final application [1]. Continuous phase or matrix is generally composed of a resin which can be thermoplastic or thermosetting polymers [2]. The composite materials with polymeric matrix are able to combine the good mechanical resistance offering the possibility to obtain materials with the desired properties [3, 4].

The final properties of a composite depend directly on the constituent phase (fiber/matrix) characteristics, the relative amount of fibers, and the geometry of dispersed phase (reinforcement). In addition to physical and morphological character-

F. P. D. Lopes (✉) · C. M. Fontes Vieira
UENF - State University of the Northern Rio de Janeiro, UENF, Advanced Materials Laboratory, LAMAV, Av. Alberto Lamego, 2000, 28013-602 Campos dos Goytacazes, Brazil
e-mail: felipeperisse@gmail.com

S. Ikhmayies et al. (eds.), *Green Materials Engineering*, The Minerals, Metals & Materials Series, https://doi.org/10.1007/978-3-030-10383-5_6

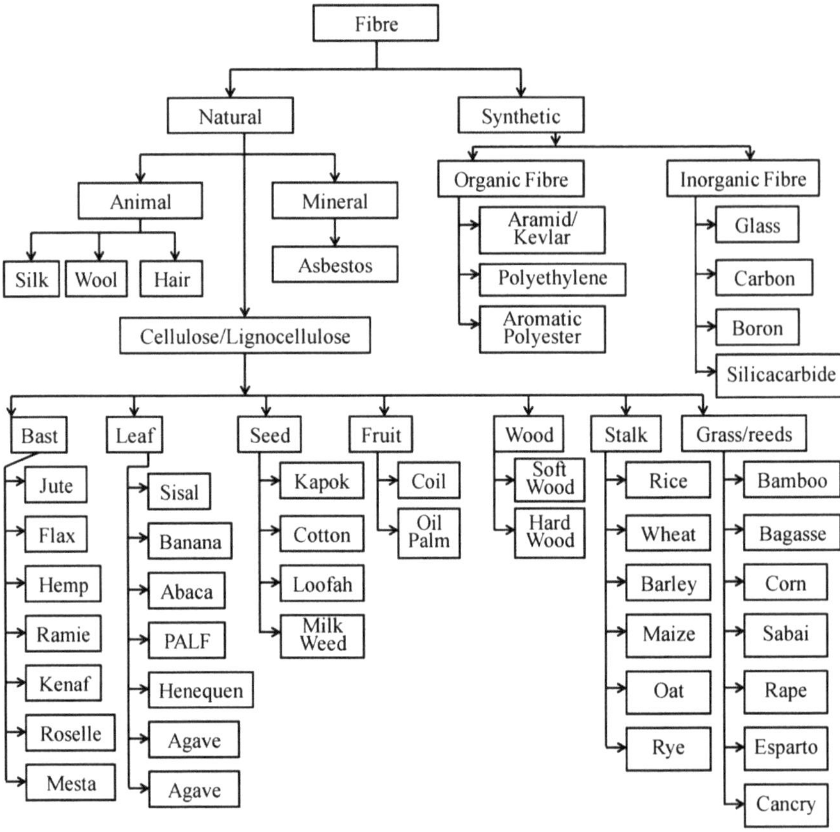

Fig. 1 Classification of natural fibers [18]

istics of constituent phases, the final composite performance depends on mechanical properties: matrix, reinforcement and of the effective fiber/matrix interaction [5].

Most of the polymeric matrices used in composites are thermosetting resins, especially epoxy, phenolic, and polyester resins. The polyester resins, in addition, to a low-cost, good dimensional stability, easy handling, also have satisfactory mechanical properties, which makes them feasible as a polymeric matrix widely used in industry [6–8]. In other hand, the epoxy resin has a higher cost, but they have electrical, thermal resistance and are good chemically; due to its high mechanical and chemical resistance, it is widely used in special laminates, pipes, tanks, aircraft, ships, and others [9–11].

As reinforcement, the use of lignocellulosic fibers is increasing; justified by its: low cost, abundance, renewability, low density, and satisfactory mechanical properties, which makes these raw materials a potential alternative of synthetic fibers, such as glass fiber, in polymeric composites [12, 13].

Figure 1 shows a natural fiber classification from where they are extracted.

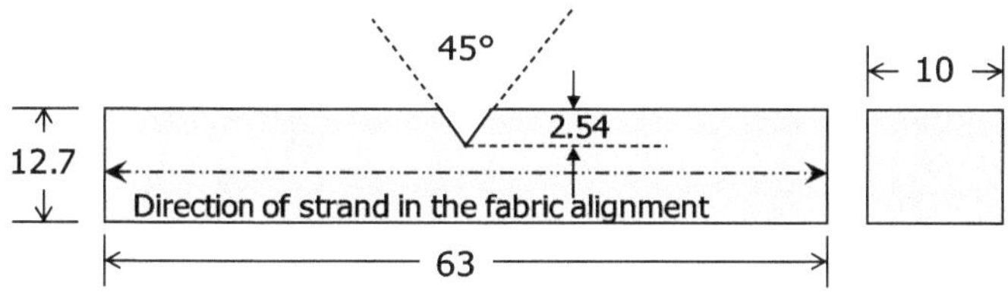

Fig. 2 ASTM D 256 Izod V-notched specimen

The impact test is the main testing to determine an energy used to deform and specimen rupture during the impact. It involves subjecting the notched specimen to a bending caused by an impact hammer. This test can be performed according to two configurations: Charpy and Izod; they are differentiated at specimens' configuration and how the specimens are attached at machine used for testing [16, 17]. According to ASTM D6110-04 [17], the Charpy impact test specimens have a square section of 10 mm besides and 120 mm in length, with a V-shaped notch in the specimens' center.

The present study proposes a comparative analysis of mechanical behavior of polyester matrix with epoxy matrix composite reinforced by 10, 20, and 30% vol of natural fibers. The specimens were tested in a PANTEC pendulum, model XC-50 located on LAMAV/UENF. The results obtained allowed to characterize these new composite materials as the impact energy required to their rupture contributing to the searching process new sustainable, viable economically and technologically advanced materials.

Methodology

To compare the values of the results from our research group were collected from ten different composites, by more than ten years of group experience in polymeric composites reinforced by lignocellulosic fibers.

The specimens are rectangular pieces of natural fibers reinforcing polymeric resins in composites. The composites were made with dimensions of 63.0 × 12.7 × 10.0 mm for Izod Testing and 125 × 12.7 × 10.0 mm for Charpy Testing according to the ASTM D 256 standard, as shown in Figs. 2 and 3.

Results

The results for the impact tests made in LAMAV by the years of research at this field of green composites are showed below, according to both tables: Table 1 for Izod tests and Table 2 for Charpy tests.

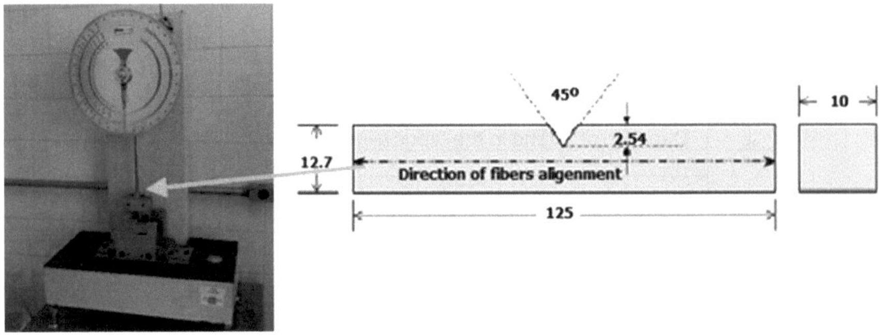

Fig. 3 ASTM D 256 Charpy V-notched specimen

Table 1 Composites tested in Izod configuration

Composite	Impact energy (Jm) by amount of fiber in %vol			Refs.
	10	20	30	
Epoxy/Fique	117	181	323	[7]
Polyester/Hemp	75	125	198	[8]
Epoxy/Piassava	75	110	165	[8]
Epoxy/Hemp	33	46	53	[9]

Table 2 Composites tested in Charpy configuration

Composite	Impact energy (Jm) by amount of fiber in %vol			Refs.
	10	20	30	
Polyester/Sisal	655	1257	1794	[10]
Epoxy/Fique	317	444	485	[11]
Epoxy/Malva and Jute	75	145	200	[12]
Polyester/Curaua	50	80	150	[13]
Epoxy/Hemp	81	94	147	[14]
Polyester/Hemp	16	39	58	[15]

Evaluating the results is easy to identify that composites made with fique fibers' reinforcing epoxy resin have very good results in both tests, comparing with the others composites analyzed. But if we look into results from sisal fiber reinforcing a polyester resin, the results from Charpy Testing are very good, but shall be confirmed in the near future doing the Izod testing.

For next steps of this research group, it is interesting to compare these results with composites made using synthetic fibers; these studies will be done soon.

Another research fronts that will be looking for news raw materials, for example, new natural fibers, bio-thermosetting resins and simulating these composites in some specific applications that requires good impact properties.

Acknowledgements The authors thank the support provided by the Brazilian agency FAPERJ.

References

1. Aquino RGMP, Almeida JRM, Monteiro SN (2005) Análise do compósito de piaçava/resina poliéster, como substituto da madeira. In: 60º Congresso Anual da Associação Brasileira de Metalurgia e Materiais- ABM, Belo Horizonte-MG, pp 1484–1492
2. Adamian R, Medina V, Weisz J (2009) Novos materiais: tecnologia e aspectos econômicos. COOPE- UFRJ, Rio de Janeiro
3. Monteiro SN, Rodrigues RJS, Souza MVE, D'Almeida JRM (1998) Sugar cane bagassewaste as reinforcement in low cost composites. Adv Performance Materials 5:183–191
4. Ventura AMFM (2009) Os Compósitos e a sua aplicação na Reabilitação de Estruturas metálicas. Ciência & Tecnologia dos Materiais 21(3/4):10–19
5. Hage E Jr (1989) Compósitos e Blendas Poliméricas. Instituto Latino Americano de Tecnologia e IBM, Campinas
6. Goodman SH (1998) Handbook of thermoset plastics, 2nd edn. Noyes Publications, New Jersey
7. Mallick PK (1993) Fiber reinforced composites, New York
8. Bledzki AK, Gassan J (1999) Composites reinforced with cellulose based fibres. Prog Polym Sci 24:221–274
9. Strong AB (1989) Fundamental of composite: materials, methods, and applications. Society of Manufacturing Engineers, Dearborn, Michigan
10. Rosato DV, Rosato DV (2004) Reinforced plastics handbook. Elsevier
11. Da Silva RV, Aquino EMF, Rodrigues LPS, Barros ARF (2008) Desenvolvimento de um compósito laminado híbrido com fibras natural e sintética. Revista Matéria 13(1):154–161
12. Pacheco EB, Santos MS, Dias ML (2000) Materiais reciclados á base de PET a cargas de coco, Brasil
13. Crônier D, Monties B, Chabbert B (2005) Structure and chemical composition of bast fibers isolated from developing hemp stem. J Agric Food Chem 53:8279–8289
14. Lavengood RE, Silver FM (1988) Interpreting supplier data sheets. Eng Mater Handb-Eng Plast Metals Park ASM Int 2:638–645
15. Souza SA (1982) Ensaios Mecânicos de Materiais Metálicos: Fundamentos Teóricos e Práticos. Edgard BlucherLtda, São Paulo
16. ASTM, International standard test methods for determining the Charpy impact resistance of notched specimens of plastic: ASTM D 6110-10, USA (2002)
17. Fu SY, Lauke B, Mäder E, Hu X, Yue CY (1999) Fracture resistance of short-glass-fiber-reinforced and short-carbon-fiber-reinforced poly-propylene under Charpy impact load and its dependence on processing. J Mater Process Technol 89(90):501–507
18. Jawaid M, Abdul Khalil HPS (2011) Cellulosic/synthetic fibre reinforced polymer hybrid composites: a review. Carbohydr Polym 86:1–18

Impact Energy Evaluation of Natural Castor Oil Polyurethane Matrix Composites Reinforced with Jute Fabric

José Gustavo de Almeida Machado, Juliana Peixoto Rufino Gazem de Carvalho, Anna Carolina Cerqueira Neves, Felipe Perissé Duarte Lopes, Sérgio Neves Monteiro and Carlos Mauricio Fontes Vieira

Abstract Motivated by the increasing concern with environmental problems, society and industry are investing materials that are ecologically friends, low-cost, and responsive to the applications they will be subjected to. In this context, the use of natural resins and the use of vegetable-based fibers are an alternative to solving these market requirements by replacing high-cost, non-organic synthetic materials with eco-friendly products. Thus, the objective of this work is to evaluate the properties of a natural matrix, produced with polyurethane based on castor oil and reinforced with jute fabric. Firstly, in order to characterize the fabric, the density of the jute was determined and the spacing between the wefts was analyzed. The materials were produced with variation of layers of fabric between 1 and 6. Subsequently, the energy absorption through the Charpy impact test was evaluated. In Conclusion, it can be emphasized that the fabric, as reinforcement, generates a significant improvement in the properties obtained.

Keywords Natural fiber · Jute fabric · Polyurethane resin · Castor oil · Impact test

Introduction

Nowadays, synthetic materials are increasingly being questioned about their use, because the concern of society regarding the preservation of the environment requires the search for new alternatives that are, at the same time, meet the specifications of the projects, low cost, and environmentally friendly. The substitution of synthetic

J. G. de Almeida Machado · J. P. R. G. de Carvalho (✉) · A. C. C. Neves · F. P. D. Lopes · C. M. Fontes Vieira
Advanced Materials Laboratory – LA-MAV, State University of the Northern Rio de Janeiro – UENF, Avenida Alberto Lamego, 2000, Campos dos Goytacazes 28013-602, Brazil
e-mail: julianarufino.arq@gmail.com

S. N. Monteiro
Materials Science Department, Military Institute of Engineering – IME, Praça General Tibúrcio, 80 Urca, Rio de Janeiro, RJ 22290-270, Brazil

S. Ikhmayies et al. (eds.), *Green Materials Engineering*, The Minerals, Metals & Materials Series, https://doi.org/10.1007/978-3-030-10383-5_7

fibers and petroleum-based resins is the current market trend and research on this area is increasingly encouraged by all stakeholders [1, 2].

Natural fibers are capable of replacing synthetic fibers, which are used in the automotive and aeronautical sectors, with advantages such as good toughness and less abrasion equipment used in composite processing. This tendency toward substitution for natural fibers is motivated because synthetic fibers are not recyclable and have a high production value, unlike natural fibers, which are abundant, renewable, recyclable, biodegradable, and essentially emission neutral of carbon dioxide, which causes the greenhouse effect [3–6].

The objective of this study was to analyze the properties of the composites with the polyurethane matrix derived from castor oil using as reinforcement the jute fabric when subjected to the Charpy impact test.

Methodology

The jute fabric was purchased commercially in a store in the city of Campos dos Goytacazes (RJ) being acquired 5 m of this fabric. It was cut into rectangles of dimensions 155 × 125 mm, washed in running water and placed in the oven for 24 h at a temperature of 60 °C until its mass was constant, thus ensuring that the fabric was completely dry.

To determine the length and width values of spacing between the fibers, three stitches of each fabric layer were measured, the total of 20 layers of fabric being used to obtain the mean values. The equipment used to carry out the measurement was the PANTEC profile projector. The measurements were performed at the Fibers Laboratory, located at LAMAV/UENF. The mean value of the fiber spacing length was (2.664 ± 0.128) mm, and the mean value of the fiber spacing width was (3.109 ± 0.123) mm. The synthesis of results is shown in Fig. 1.

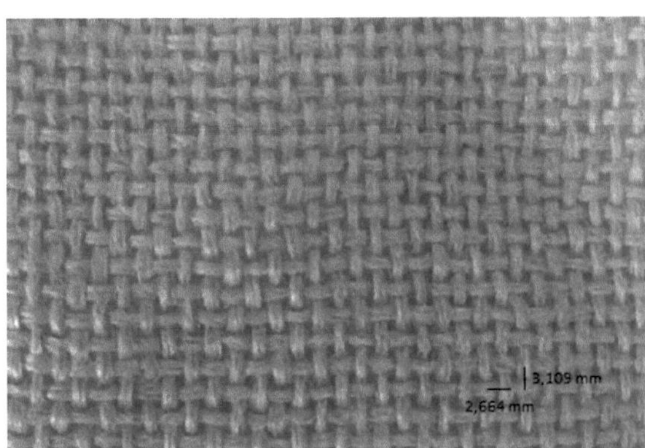

Fig. 1 Jute fabric with fiber spacing dimensions

Density Determination of Jute Fabric

The density calculation in relation to the distilled water was carried out using a pycnometer and a digital scale with an accuracy of ±0.001 g. First, the tare of the digital scale was made with the empty pycnometer, obtained the mass of the pycnometer + water. After that the mass of the pycnometer/solid/distilled water was measured, thus obtaining the apparent density of the fiber in relation to the water. Before placing the fabric in the pycnometer with distilled water, it was kept under vacuum for 2 h in order to remove the existing bubbles. Parallel to the test, the moisture correction factor that might exist in the material was calculated. To this end, a quantity of fabric was weighed and placed in a greenhouse for 24 h, and then cooling was done under vacuum in order to avoid absorbing the moisture from the medium. The expression for calculating the density (ds) of the fabric as a function of the water density was given by Eqs. 1 and 2, where m_1 is the mass of the empty pycnometer; m_2 is the mass of the pycnometer with the fiber; m_3 is mass of the pycnometer with fiber + water; and m_4 is mass of the pycnometer with water;

$$ds = \frac{\rho \text{s}}{\rho \text{ water}} = \frac{\text{m2} - \text{m1}}{(\text{m4} - \text{m1}) - (\text{m3} - \text{m2})} \quad (1)$$

$$\rho \text{s} = ds.\rho \text{ water} \quad (2)$$

As the scale was calibrated with the pycnometer, the calculation was performed considering that the mass of the empty pycnometer was zero. A 50 ml pycnometer was used to raise the density.

Composites Produced with Castor Oil Resin

For the preparation of the test bodies, the volume matrix 186 cm^3 was used, suitable for molding the specimen for the Charpy impact tests. The binder used was the polyurethane bicomponent resin derived from manganese oil, provided by Imperveg. The proportions, among the components A and B, adopted were 1:1.5 by mass.

The amount of jute fabric layers constituting the test bodies ranged from 1 to 6 layers. The composites produced were subjected to a cold pressure of 2 MPa in a hydraulic press for 24 h, and after that time, the body was removed from the mold and left to complete the cure under ambient conditions for 7 days. After the deadline, the samples were placed in an oven for 48 h at 60 °C.

For each plate produced, at least 10 specimens were cut for the Charpy impact test, measuring 125 × 12.7 × 10 mm, in accordance with ASTM D6110-18.

Table 1 Density of fibers [2]

	Cotton	Linen	Sisal	Carbon	Aramida	E-glass	S-glass
Density (g/cm^3)	1.5–1.6	1.5	1.5	1.4	1.4	2.5	2.5

Table 2 Energy absorbed in Charpy impact test of composites with castor oil matrix reinforced with jute fabric composites

	Number of layers					
	1	2	3	4	5	6
Energy absorbed (J/m)	222.60	130.15	87.97	100.17	131.77	194.26

Results and Discussion

After density testing using a pycnometer with water, the density value of the jute fabric obtained was 0.67 g/cm^3. This value was used to determine the resin/reinforcement ratio in the sample preparation for the Charpy impact test. Table 1 shows the results of the densities of other fibers found in the literature [2]. From this, it is possible to conclude that jute is considered a low-density fiber, when compared to synthetic fibers, and its density is more than three times smaller when compared to E-glass and S-glass. This is an important factor for the production of lightweight composites.

Figure 2 shows the impact energy variation of the natural matrix test bodies varying the number of layers from 1 to 6 of jute fabric. It is observed that the composites of 1 and 2 layers exhibited an unexpected behavior, due to the notched tenacity of the composite with 1 layer being significantly larger than the composites with 2 layers and more. According to the literature [2, 7], the expected behavior was the reverse. This distortion in the values of the specimens from 2 layers can be attributed to problems in the processing of the specimens. It also impaired the use of the fabric as a reinforcement of the polymer matrix and prevented a greater energy absorption before the samples were broken. In parallel, 1 layer was already sufficient to promote a good energy absorption, which can be verified comparing with the literature results [8].

The composites of 3–6 layers of jute fabric showed an increase of the impact energy, which was already expected, since the increase of jute fabric in the polymer matrix increases the toughness to the notch [9].

It is important to note that the 3–6 layers (Fig. 3) test specimens did not break completely after the test because the jute fabric acted as if it "held" the polyethylene matrix, not allowing the composite to completely rupture [8].

Table 2 shows the variation of the results of the absorbed energies according to the number of layers. It is apparent that from 3 layers of fabric an increased energy absorption has been achieved. In other words, the energy absorption increased as new layers of fabric were added.

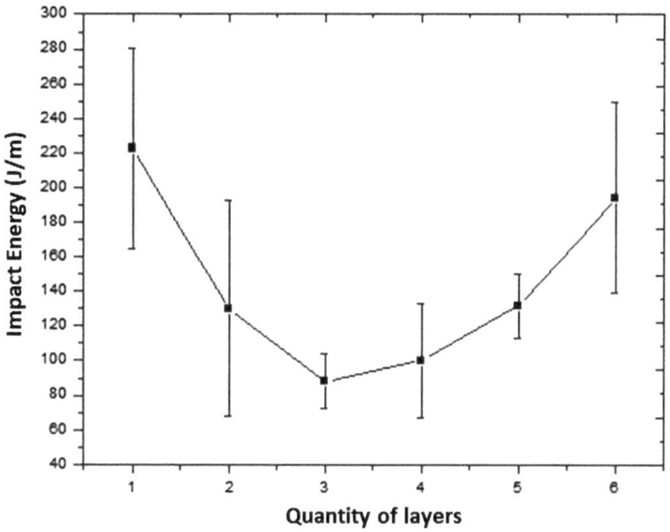

Fig. 2 Graphic of the variation of the impact energy absorbed correlated to the number of layers of jute fabric

Fig. 3 Samples after impact test

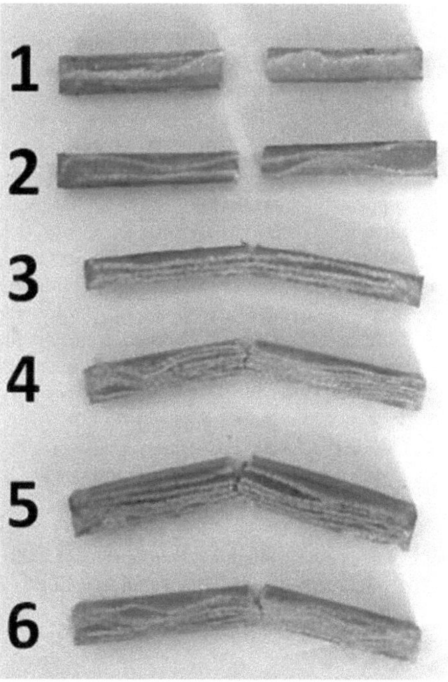

Conclusions

Jute fabric has a considerably light density compared to other natural and synthetic fibers. The samples reinforced with 3–6 layers of jute fabric showed a growth of the energy absorbed by the impact, ranging from 87.97 to 194.26 J/m, which was already expected, since the increase in the amount of reinforcement of fiber, increases the toughness to the notch of the composites [10].

The composites with these amounts of jute fabric layers did not completely rupture after the Charpy impact test because the jute fabric "held" the polyurethane matrix so that it did not completely rupture [8]. In general, eco-friendly polymer composites showed a good performance when subjected to impact in relation to other results found in the literature, even when compared with synthetic matrices.

Acknowledgements The authors would like to thank FAPERJ (E-26/202.128/2017), Imperveg for the resin, the Laboratory of soils (LSOL), located at CCTA/UENF, the Laboratory of Advanced Materials—LAMAV UENF for the support and, mainly to Renan da Silva Guimarães.

References

1. Bledzki AK, Gassan J (1999) Composites reinforced with cellulose based fibers. Prog Polym Sci 24:221–274. https://doi.org/10.1016/s0079-6700(98)00018-5
2. Callister WD Jr (2000) Materials science and engineering—an introduction, 5th edn. Wiley, New York
3. Camerini AL, Terrones LAH, Monteiro SN (2008) Tenacidade ao impacto de compósitos de tecido de juta reforçando matriz de polietileno reciclado. Revista Matéria. http://www.materia.coppe.ufrj.br/sarra/artigos/artigo10961/. Accessed 21 June 2018
4. Silva RV (2003) Compósitos de Resina Poliuretano derivada de Óleo de mamona e fibras vegetais. PhD thesis, São Paulo University
5. Gon D, Das K, Paul P, Maity S (2012) Jute composites as wood substitute, international journal of textile. Science 1(6):84–93. https://doi.org/10.5923/j.textile.20120106.05
6. Gore A (2006) An inconvenient truth. the planetary emergency of global warming and what we can do about it. Rodale Press, Emmaus, Pennsylvania, USA
7. Sahed ND, Jog PP (1999) Natural fiber polymer composites: a review. Adv Polym Technol 18:221–274
8. Camerini AL (2008) Caracterização e propriedades de compósitos de tecido de juta reforçando matriz de polietileno reciclado. PhD thesis, State University of the Northern Rio de Janeiro
9. Monteiro SN, Lopes FPD, Ferreira AS et al (2009) JOM 61:17. https://doi.org/10.1007/s11837-009-0004-z
10. Mohanty AK, Misra M, Hinrichsen G (2000) Biofibers, biodegradable polymers and biocomposites: an overview. Macromol Mat Eng 1–24:276–277

Comparison of Interfacial Adhesion Between Polyester and Epoxy Matrix Composites Reinforced with Fique Natural Fiber

Michelle Souza Oliveira, Artur Camposo Pereira, Fabio da Costa Garcia Filho, Luana Cristyne da Cruz Demosthenes, Larissa Fernandes Nunes, Fabio de Oliveira Braga, Fernanda Santos da Luz and Sergio Neves Monteiro

Abstract Polyester and epoxy resins are thermosetting polymers extensively used as composite materials matrices. The advantage of the polyester resin in comparison with the epoxy matrix is its lower cost. On the other hand, its permeability can be considered a disadvantage, once as the solvents dry the resin becomes more porous. This represents a direct impact on its mechanical properties. Epoxy resin has no solvents and its polymerization occurs only when two components are mixed, resin and hardener. The processing of this type of polymer requires control of temperature and humidity during its production. Polyester resin continues to cure over time, are prone to cracking and breaking, while epoxy resin once cured retains its full properties. The current pullout tests were performed to compare the interfacial adhesion of fique fibers with epoxy and polyester polymer matrix. The results indicated a critical length approximately 65% higher for the fique fiber/polyester in comparison to fique fiber/epoxy as well as an interfacial strength 4.9 times higher, which may indicate stronger adhesion of fique fiber with epoxy resin.

Keywords Mechanical performance · Pullout test · Fique fiber

Introduction

The use of green materials for engineering applications has gained high attention from industries in recent years due to potential to reduce waste from non-degradable

---

M. S. Oliveira (✉) · A. C. Pereira · F. da Costa Garcia Filho · L. C. da Cruz Demosthenes · L. F. Nunes · F. S. da Luz · S. N. Monteiro
Military Institute of Engineering—IME, Rio de Janeiro, Brazil
e-mail: Oliveirasmichelle@gmail.com

F. de Oliveira Braga
National Service of Industrial Apprenticeship—SENAI, Rio de Janeiro, Brazil

F. de Oliveira Braga
Fluminense Federal University—UFF, Rio de Janeiro, Brazil

S. Ikhmayies et al. (eds.), *Green Materials Engineering*, The Minerals, Metals & Materials Series, https://doi.org/10.1007/978-3-030-10383-5_8

synthetic materials and their carbon footprint [1–3]. Among these green materials, the use of natural fibers as reinforcement of polymer matrix composites is considered promising candidates for engineering applications in automobilist industry and even for ballistic armour systems [4]. In spite of numerous publications dedicated to natural fiber composites, the growing interest for engineering applications is continuously demanding research works on less common but promising natural fibers. An example is the relatively unknown Fique fiber (*Furcraea andina*), which is extracted from the leaves of an Andean plant. Fique fiber has brought attention due to its mechanical properties, which qualifies this fiber for potential use as a composite reinforcement [4–7]. Table 1 presents some of the mechanical properties that have been reported for the fique fiber [5–7].

Studies of interface modification and characterization of natural fiber reinforced polymer matrix composites revealed that the quality of the fiber-matrix interface is essential for the application of natural fibers as reinforcement for polymeric materials. Also, it has been showed that the different nature of the fibers and matrices have a direct impair in the properties of the composite. Therefore, strong adhesion of interfaces is essential to achieve successful transfer and distribution of stress through the material [8–15]. The focus of this work was the mechanical interaction between the fique fiber and epoxy resin matrix as well as the fique fiber and polyester resin matrix, once these two thermosetting polymers are extensively used as composite matrices.

Epoxy resins, also known as poly epoxides, are a class of reactive pre-polymers and polymers, which contain epoxide groups. Epoxy resins may be reacted (cross-linked) either with themselves through catalytic homo polymerization, or with a wide range of co-reactants. Reaction of poly-epoxides with themselves or with poly-functional hardeners forms a thermosetting polymer, often with high mechanical properties, and temperature and chemical resistance. Epoxy has a wide range of applications, including metal coatings, use in electronics/electrical components, high-tension electrical insulators, fiber-reinforced plastic materials, and structural adhesives [16].

On the other hand, unsaturated polyesters are condensation polymers formed by the reaction of polyols (also known as polyhydric alcohols), organic compounds with multiple alcohol or hydroxy functional groups, with saturated or unsaturated dibasic acids. The liquid resin is converted to a solid by cross-linking chains. This is done by creating free radicals at unsaturated bonds, which propagate in a chain reaction to other unsaturated bonds in adjacent molecules, linking them in the process. The free radicals are induced by adding a compound that easily decomposes into free radicals. Polyester resins are thermosetting and, as with other resins, cure exothermically [17].

Table 1 Mechanical properties of the fique fibers [5–7]

Tensile strength (MPa)	237 ± 51
Strain at break (%)	6.02 ± 0.69
Tensile modulus (GPa)	8.01 ± 1.47
Average diameter (mm)	0.16

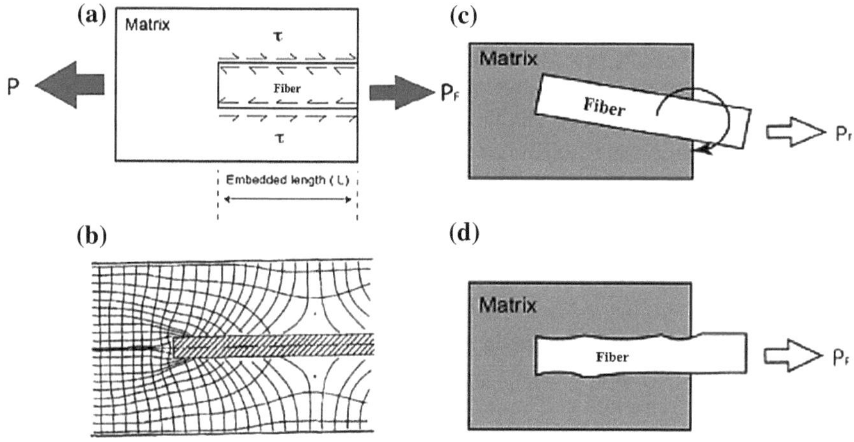

Fig. 1 **a** Interfacial bonding stress, **b** tension at the ends of a fractured fiber, **c** and **d** shape and direction of fiber [9]

Thereby, the interfacial bonding stress in the fiber pullout model is influenced by several parameters such as:

- Fiber embedded length (l);
- Young's modulus of fiber (E_f) and matrix (E_m);
- Thickness of adhesive layer (t_a) and;
- The diameter (D_f) of the fiber, which represents the total bonding area on the fiber surface.

The mechanical properties of natural fibers may vary due to their porous structures. Thus, the fiber pull-out model is only able to provide a rough prediction as the mechanical interaction between the fiber and matrix is not taken into account in this model [9]. This model is illustrated in Fig. 1a and b. It is necessary to point out that in a real scenario, natural fibers are not aligned in one direction (Fig. 1c) and their surface varies along its length (Fig. 1d).

Materials and Methods

For this investigation, first it was necessary to measure the diameter of the fibers, which were received as bundles. The procedure consisted of randomly selecting one hundred fibers and measuring their equivalent diameters, which corresponds to the average between the larger and smaller (90° rotation) cross section dimensions at five different locations along the length of each fiber. As a whole, the main concentration of diameter values considered in this work reflected those reported by Teles et al. [6], who found that most mechanically extracted fique fibers have a diameter between 150 and 180 microns.

The fibers were dried at 60 °C in an air-oven for 24 h, in order to reduce the inherent moisture of the fibers. The epoxy resin used as the embedding medium was commercial epoxy type diglycidyl ether of bisphenol A (DGEBA) hardener with triethylenetetramine (TETA). The DGEBA/TETA epoxy with stoichiometric phr 13 amount of hardener was cured at room temperature. As also, an isophtalic polyester resin hardened with 1% methyl-ethyl-ketone, was used as the composite matrix.

Pullout tests were performed using embedded fiber lengths, L, varying from 3, 5, 10, and 15 to 20 mm. The single fibers were mounted on epoxy and polyester sockets 40 mm long. The specimens were subjected to a tensile force using universal testing machine and the load–displacement curve was recorded. The free fiber end was clamped with pneumatic action grips. A cross-head speed of 1.0 mm/min was used in all the experiments.

The interfacial strength (τ_c) could be calculated using the following equation:

$$l_c = \frac{d\sigma_f}{2\tau_c} \quad (1)$$

Where σ_f is the tensile strength of the fiber and d is the equivalent diameter of the fiber. The value of d can be a source of error for natural fibers, since their cross sections are not usually circular, and can vary in forms and dimensions over fairly large bounds.

Equation (1) provides a value of interfacial shear strength that is apparent because it depends on both the embedded length and the diameter of the fiber, which are usually assumed constant [14].

Depending on the degree of the fiber/matrix adhesion, such tensile forces are transferred from the matrix to the fiber. These tensile stresses in the fiber continue to grow until the fiber tensile strength is reached and the fiber fractures at some point where the stress concentration is high enough. This loading process continues until the fiber-fragment lengths are so small that the tensile stresses induced in the fiber can no longer reach the fiber tensile strength [9, 10, 14].

Results and Discussion

The result typically expected for this kind of test is the fiber being pulled out of the matrix without fracturing, meaning that adhesive failure rather than cohesive failure occurs.

Figure 2 exhibits two examples of the typical load versus displacement curves obtained in this work.

The load versus displacement curves are used to determine the shear strength at the interface between the fiber and the matrix using the Kelly and Tyson [15] equation, as suggested by Monteiro and D'almeida [12].

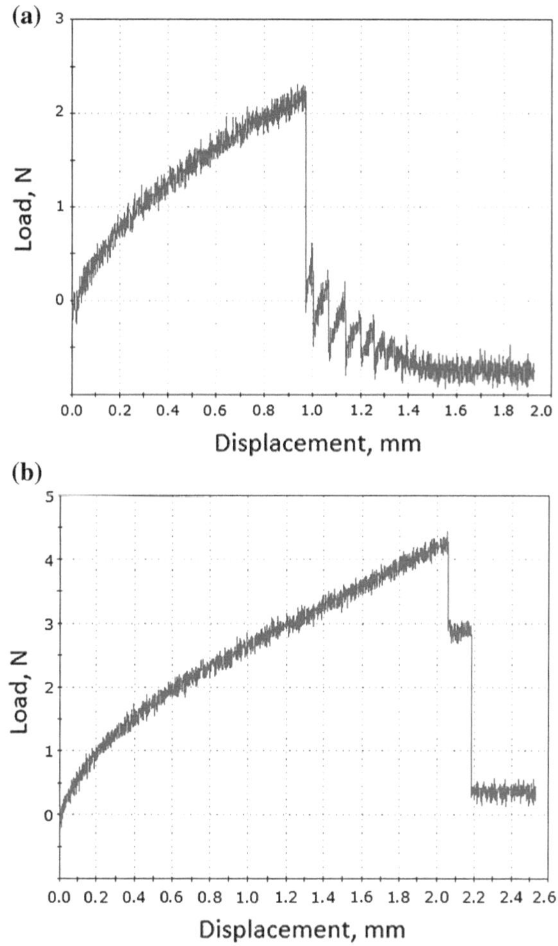

Fig. 2 Examples of the typically registered load versus extension curves for fique fiber: **a** epoxy and **b** polyester matrix

Table 2 Parameters for fique/epoxy and fique/polyester

	Fique/epoxy	Fique/polyester
First linear section	$\sigma = 0.44L + 8.37$	$\sigma = 9.03L + 30.18$
Second linear section	$\sigma = 5.98L - 8.37$	$\sigma = 2.16L + 62.34$
Critical length	3.02 mm	4.68 mm

Figure 3 shows the pullout curve for fique fibers embedded in epoxy (Fig. 3a) and polyester (Fig. 3b) capsules. Table 2 summarizes the test parameters for the fique/epoxy system as well as the fique/polyester systems.

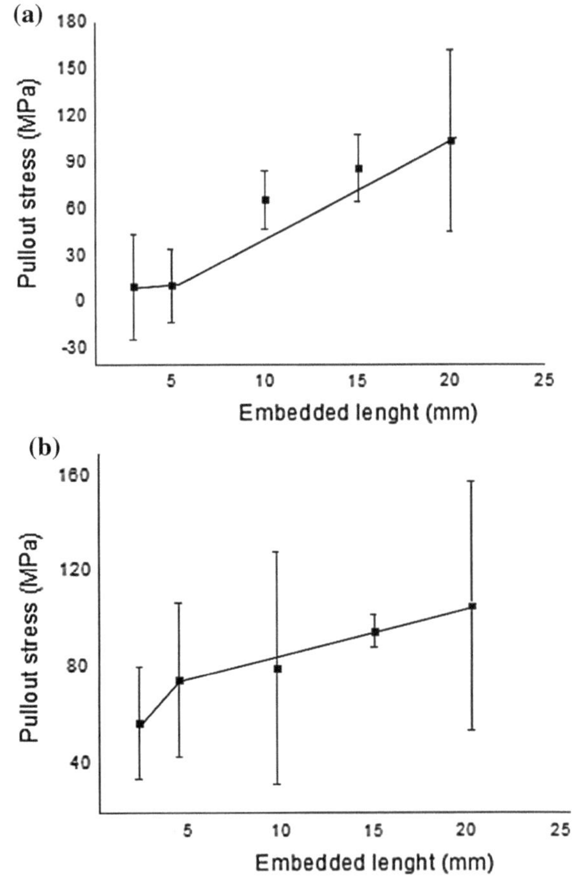

Fig. 3 Pullout test of fique fibers: stress *versus* embedded length curve **a** epoxy and **b** polyester matrix

Since small values of l_c indicate greater interfacial fiber/matrix adhesion [11], these results of approximately 65% higher critical length reveals better adhesion between the system fique/epoxy in comparison with fique/polyester.

The interfacial strength (τ_c) of fique/epoxy and fique/polyester was evaluated by the equation of Kelly and Tyson [15] and presented in the Table 3.

Table 3 Critical fiber pullout length and interface shear strength of fique fibers

Fiber	l_c (mm)	τ (MPa)	Resin matrix	Ref.
Fique	3.02	0.27	Epoxy	PW
Fique	4.68	1.32	Polyester	PW
Coir	12.4	1.42	Epoxy	[10]
Coir	5.38	0.43	Polyester	[11]
PALF	7.30	4.93	Epoxy	[10]
Curauá	10.20	0.49	Polyester	[11]

PW = Present Work

(a) (b)

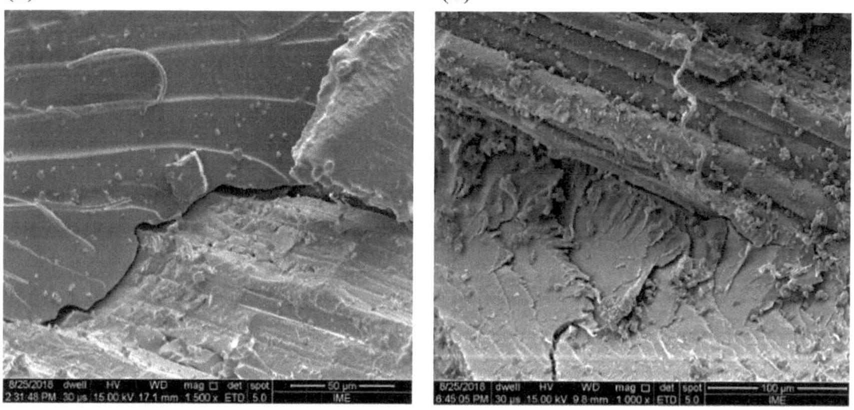

Fig. 4 SEM of the interface between fique fiber and **a** epoxy, and **b** polyester matrix

Comparing the interfacial strength calculated for the two conditions tested with other values reported on the literature, one may notice that the fique/polyester interface is weaker than fique/epoxy. Figure 4 shows the interface between the fique fiber and thermoset matrices.

Summary and Conclusions

The fique fiber presented in pullout tests a critical length about 65% higher in polyester than epoxy, which indicates weaker interfacial adhesion between this fiber and polyester than epoxy matrix. In addition, the value of interfacial strength for the fique/polyester was 4.9 times greater than fique/epoxy.

Acknowledgements The authors thank the Brazilian agencies CAPES and CNPq for the financial support, and CAEX for performing the ballistic tests.

References

1. Ramesh AV et al (2018) Comparitive study of the mechanical properties of alkali treated and untreated sugarcane bagasse fiber reinforced composite material. Int Res J Eng Technol 5(5):413–420
2. Valadez-Gonzalez A et al (1999) Effect of fiber surface treatment on the fiber–matrix bond strength of natural fiber reinforced composites. Compos Part B Eng, 309–320
3. Herrera-Franco PJ, Valadez-Gonzalez A (2005) A study of the mechanical properties of short natural-fiber reinforced composites. Compos Part B Eng 36:597–608
4. Monteiro SN et al (2018) Fique fabric: a promising reinforcement for polymer composites. Polymers 10(246):10
5. Gañán P, Mondragon I (2002) Surface modification of fique fibers. Effects on their physico-mechanical properties. Polym Compos, 383–394
6. Teles MCA et al (2015) Fique fiber tensile elastic modulus dependence with diameter using the Weibull statistical analysis. Mater Res 18(Suppl 2):193–199
7. Netto PA et al (2016) Correlation between the density and the diameter of fique fibers. Mater Sci Forum 869:377–383
8. Aquino RCM, Monteiro SN, D'almeida JRM (2003) Evaluation of the critical fiber length of piassava (Attalea Funifera) fibers using the pullout test. J Mater Sci Lett 22:1495–1497. https://doi.org/10.1023/a:1026147013294
9. Lau K-T et al (2018) Properties of natural fibre composites for structural engineering applications. Compos Part B 136:222–233
10. Luz FSD et al (2018) Critical length and interfacial strength of PALF and coir fiber incorporated in epoxy resin matrix. J Mater Res Technol, 456
11. Monteiro SN, Aquino RCMP, Lopes FPD (2008) Performance of curaua fibers in pullout tests. J Matter Sci 43:489–493. https://doi.org/10.1007/S10853-007-1874-5
12. Monteiro SN, D'almeida JRM (2006) Ensaios De Pullout Em Fibras Lignocelulósicas - Uma Metodologia De Análise. Revista Matéria 11(3):189–196
13. Sydenstricker THD, Mochnaz S, Amico SC (2003) Pull-out and other evaluations in sisal-reinforced polyester biocomposites. Polym Test 22:375–380. https://doi.org/10.1016/S0142-9418(02)00116-2
14. Viel Q et al (2018) Interfacial characterization by pull-out test of bamboo fibers embedded in poly(Lactic Acid). Fibers 6(7). https://doi.org/10.3390/fib6010007
15. Kelly A, Tyson W (1965) High strength materials. Wiley, New York
16. D'almeida JR, Monteiro SN (1996) The effect of the Resin/Hardener ratio on the compressive behavior of an epoxy system. Polym Test 15:329–339
17. Brydson J (1999) Plastic materials, 7th edn. Butterworth Heinemann, Oxford

Part III
Nano and Micro Green Composites

Application of Natural Nanoparticles in Polymeric Blend of HMSPP/SEBS for Biocide Activity

Luiz Gustavo Hiroki Komatsu, Angelica Tamiao Zafalon, Vinicius Juvino Santos, Nilton Lincopan, Vijaya Kumar Rangari and D. F. Parra

Abstract The natural nanoparticles of $CaCO_3$, from eggshell and $CaCO_3$/Ag, were investigated as candidates for biocide nanoparticles in a blend of thermoplastic elastomer, styrene–ethylene/butadiene–styrene (SEBS), and polyolefin, high-melt-strength polypropylene (HMSPP). The nanoparticles with silver (Ag) were synthesized by metal precursor method, and before the application in the polymer matrix, the nanoparticles were analyzed on dynamic lighting scattering (DLS) and Raman spectroscopy. After DLS and Raman analyses, the nanoparticles were evaluated in biocide tests against *P. aeruginosa* and *S. auerus*. The aim of this investigation was the blending and the injection molding in dumbbell samples with the nanoparticles. The characterizations were carried out by differential scanning calorimetry (DSC), thermogravimetry analysis, mechanical tests, X-ray diffraction, Raman spectroscopy, scanning electron microscopy, and energy-dispersive scanning and biocide tests. Correlation between size and particle distribution on the polymer was founded. However, samples at higher concentrations (>1 wt%) do not show biocide activity.

Keywords Natural nanoparticles · Eggshell · Biocide activity

The natural nanoparticles of $CaCO_3$, from eggshell and $CaCO_3$/Ag, were investigated as candidates for biocide nanoparticles in blend of thermoplastic elastomer, styrene–ethylene–butadiene–styrene (SEBS), polyolefin, and high-melt-strength

L. G. H. Komatsu (✉) · A. T. Zafalon · V. J. Santos · D. F. Parra
Nuclear and Research Institute, IPEN-CNEN/SP, Av. Prof Lineu Prestes, 2242, Cidade Universitária, São Paulo CEP 05508-000, Brazil
e-mail: luizkomatsu@gmail.com

N. Lincopan
Department of Microbiology, Institute of Biomedical Sciences, University of Sao Paulo, Sao Paulo CEP 05508-000, Brazil

N. Lincopan
Department of Clinical Analysis, School of Pharmacy, University of Sao Paulo, Sao Paulo, Brazil

V. K. Rangari
Center for Advanced Materials Science and Engineering, Tuskegee University, Tuskegee, AL 36088, USA

S. Ikhmayies et al. (eds.), *Green Materials Engineering*, The Minerals, Metals & Materials Series, https://doi.org/10.1007/978-3-030-10383-5_9

polypropylene (HMSPP). The nanoparticles with silver (Ag) were synthesized by metal precursor method, and before the application in the polymer matrix, the nanoparticles were analyzed on dynamic lighting scattering (DLS) for size determination and zeta potential of the nanoparticles. After DLS, the nanoparticles were evaluated in biocide tests against *P. aeruginosa* (ATCC 27853) and *S. auerus* (ATCC 6538p). The aim of this investigation was to prepare the blending and the injection molding in dumbbell samples with the nanoparticles. The characterizations were carried out by differential scanning calorimetry (DSC), thermogravimetry analysis (TGA), mechanical tests, X-ray diffraction (DRX), scanning electron microscopy (SEM), and energy-dispersive scanning (EDS). For the biocide activity, tests were compared, with the same bacteria and methodology, the $CaCO_3$ and $CaCO_3$/Ag nanoparticles. It was founded that there is a correlation between size and particle distribution on the polymer. However, samples at higher concentrations (>1% wt%) do not show biocide activity.

Introduction

Considering the environmental risks of the synthesized nanoparticles and the reagents used, eco-friendly chemistry approaches are appreciated to avoid possible environmental contamination. Biologically derived metallic carrier nanocomposites are garnering considerable attention and are widely being implemented in several biomedical and engineering applications due to their potent physicochemical properties [1].

Eggshell is a biologically derived metallic carrier, utilized for this study. Generally, egg shell is a reinforcing component in structural materials that helps in mitigating challenges while providing other societal needs. Few technologies are available for the processing of eggshells into useful products. More innovations in recovery technologies for turning eggshells into resource materials are currently need to convert such natural and rich source of $CaCO_3$ into other forms that are useful for functional and sustainable material design. Past findings show that eggshell contains about 95% $CaCO_3$ which can be used to modify the weak properties of polymers [2].

Is reported in the literature [2] that eggshell with silver (Ag) has a biocide activity against some types of bacteria [2]. The antimicrobial activity of silver materials (metallic silver and the silver (I) ion) has been known for years. Since ancient times, human beings have used silver empirically as dishes and as a water preservative. The most widely documented uses are prophylactic treatment of burns and water disinfection. The toxicity of silver materials to human cells is considerably lower than that against bacteria [3].

It has been reported recently that silver in the form of nanoparticles (AgNPs), especially for medical applications, is a promising alternative to silver salts and bulk metal. Salts may release the silver quickly and uncontrollably, while bulk metal is too inefficient releasing material. AgNPs exhibit outstanding physical properties that are obviously different from those of ions and bulk metal, showing promise of their use in many applications in the fields of medicine, microbiology, and analytical chemistry, among others. More importantly, AgNPs are highly antimicrobial due to their antiseptic properties against several species of bacteria, including the common kitchen microbes. As such, AgNPs have caught the attention of many researches, especially because of their extraordinary antimicrobial activity [4].

Antimicrobial susceptibility testing methods are divided into types based on the principle applied in each system. The Kirby–Bauer and Stokes methods are usually used for antimicrobial susceptibility testing, with the Kirby–Bauer method being recommended by the National Committee for Clinical Laboratory Standards (NCCLS) (NCCLS, 03). The antibacterial characteristics of silver nanoparticles produced have been demonstrated by the direct exposition of bacteria to colloid silver particle solution [5, 6].

For this application, the $CaCO_3$ and $CaCO_3$/AgNp were synthesized and melt processed in the blend of high-melt-strength polypropylene (HMSPP), obtained by gamma irradiation and styrene–ethylene–butadiene–styrene (SEBS). The key point was to evaluate the biocide effect of the nanoparticles' blends after processing.

Experimental Procedure

Extraction of Calcium Carbonate from Eggshells

The eggshells were boiled in distilled water for 4 h and dried at 50 °C until constant weight. The eggshells were mechanically triturated using ball milling for 8 h. The sample was washed with distilled water (three times) and (one time) with alcohol and centrifuged I the sequence. The supernatant was discarded, and the calcium carbonate sediment was dried at 60 °C. The calcium carbonate was characterized by dynamic lighting scattering (DLS).

Synthesis of $CaCO_3$/Ag Nanoparticles

The eggshell powder was prepared with the $AgNO_3$(1/1), and ethylenoglicol was ball-milled for 8 h. The sample was washed with distilled water three times to remove chemical surplus and centrifuged and dried in a desiccator. The $CaCO_3$/AgNp was characterized by dynamic lighting scattering (DLS).

Melt Processing of the Samples

To obtain the blend of high-melt-strength polypropylene and styrene–ethylene–butadiene–styrene (HMSPP/SEBS), the sample of isotactic polypropylene

(H603—Braskem) was placed in plastic container and irradiated by (^{60}Co) in acetylene atmosphere in order to obtain the HMSPP. The sample of SEBS (KRATON—G1633) was mixed with paraffinic oil from Nynas (Nyflex 3030), antioxidant IRGANOX 1010 from BASF, and PP-*graft*-MA from Additivant. The melt processing of HMSPP/SEBS blend was made in single-screw extruder—Thermo Haake Polymer, in the range of temperature of 160–210 °C. The extruded material was divided and mixed with different concentrations of 0.3 and 1 wt% of $CaCO_3$ and 0.3 and 1 wt% of $CaCO_3$/Ag. Injection molding in the form of dumbbell samples utilized the AX Plastics equipment, and the process was in the range of 180–210 °C.

Dynamic Lighting Scattering (DLS)

The analysis of DLS and potential zeta was carried out in equipment of Zetasizer Nano Zs from Malvern. For measurements, the samples were dispersed in distilled water and sonicated.

Biocide Tests

This test was carried out to evaluate the results of diffusion of the nanoparticles before and after the melt processing.

The bacteria utilized for this test were the *P. aeruginosa* (Gram negative) (ATCC 25923e) and *S. aureus* (Gram positive) (ATCC 6538p). With a sterile swab, the inoculums of the bacteria were seeding in all directions in sterile Petri dish with Müeller–Hinton agar; after drying, the samples were placed in contact to the agar and incubated for 24 h.

Differential Scanning Calorimetry (DSC)

Thermal properties of specimens were obtained using a differential scanning calorimeter (DSC) 822, Mettler Toledo. The thermal behavior of the films was monitored in the program: (1) heating from 25 to 280 °C at a heating rate of 10 °C min^{-1} under nitrogen atmosphere, (2) holding for 5 min at 280 °C, and (3) then cooling to 25 °C and reheating to 280 at 10 °C min^{-1}, according to ASTM D 3418-08. The crystallinity was calculated according to the equation:

$$X_C = P \times \frac{\Delta H_f \times 100}{\Delta H_0} \tag{1}$$

where ΔH_f is melting enthalpy of the sample, ΔH_0 is melting enthalpy of the 100% crystalline PP which is assumed to be 209 kJ kg^{-1}.

Thermogravimetry Analysis (TGA)

The TGA analysis was carried out in Mettler Toledo TGA/851 equipment, in heating program of 25–650 °C and heat rate of 10 °C min^{-1} under N_2 (50 mL min^{-1}) atmosphere.

Mechanical Tests—Stress and Strain

The samples were tested in a universal testing machine EMICDL3000 model with strain rate of 2.10^{-2} s^{-1}.

X-Ray Diffraction (XRD)

X-ray diffraction measurements were carried out in the reflection mode on a Rigaku diffractometer Mini Flex II (Tokyo, Japan) operated at 30 kV voltage and a current of 15 mA with CuK_α radiation ($\lambda = 1.541841$ Å).

Scanning Electron Microscopy (SEM)–Energy-Dispersive Scanning (EDS)

Scanning electron microscopy was done using an EDAX PHILIPS XL 30. In this study, a thin coat of gold was sputter-coated onto the samples.

Results and Discussion

From DLS analysis, the particles of $CaCO_3$ and $CaCO_3$/Ag showed the particle size of 1795.5 and 1752.3 nm, respectively. The micrometric size of the synthesized particles can be justified by the time used to process then in the ball milling. $CaCO_3$/Ag showed a particle size approximately equal to neat $CaCO_3$ particle. Is expected that Ag nanoparticles were formed on the surface of $CaCO_3$ due to porosity of the surface. It was not possible to detect the Ag by XRD analysis, as shown in Fig. 1.

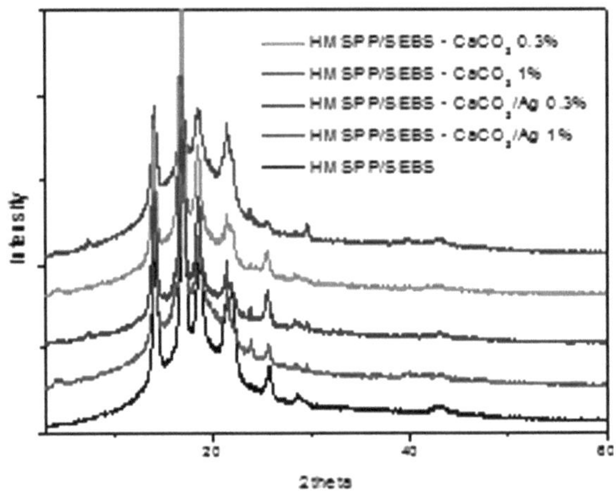

Fig. 1 DRX diffratograms of injection molded samples

Table 1 Stress and strain of the samples with $CaCO_3$ and $CaCO_3$/Ag

Sample	Tension (MPa)	Elongation (%)
HMSPP/SEBS	27.04	295.13
HMSPP/SEBS 0.3% $CaCO_3$	25.34	278.03
HMSPP/SEBS 1% $CaCO_3$	27.42	293.31
HMSPP/SEBS 0.3% $CaCO_3$/Ag	24.03	245.40
HMSPP/SEBS 1% $CaCO_3$/Ag	25.42	253.30

Nevertheless, it was possible to see the evidence of presence of Ag particles through the SEM images and EDS analysis as shown in Fig. 2.

By observing the analysis of EDS, the presence of Ag° on the neat material was very clear. The synthesized powder, however, after processing the silver was coated by the polymer. It is interesting to observe points in white color. These points are evidence of clusters formation of $CaCO_3$/Ag° on the material.

These clusters have significant influence, on the mechanical properties as shown in Table 1, with loss of modulus and elongation.

The influence of $CaCO_3$ in the blend is also observed owing to the formation of clusters that also cause a decrease in the same properties. Formation of clusters points interferes to break, and the tension resistance tends to decrease.

The influence of the particles can be observed on thermal values, as shown in Table 2.

The particles like $CaCO_3$Ag or $CaCO_3$ can act on the polymers as nucleant agent, increasing the crystallinity of material. However, this effect was not possible to observe in the present case. On the other hand, it was possible to observe the formation of the two T_{onset} on TGA, indicating the presence of each polymer.

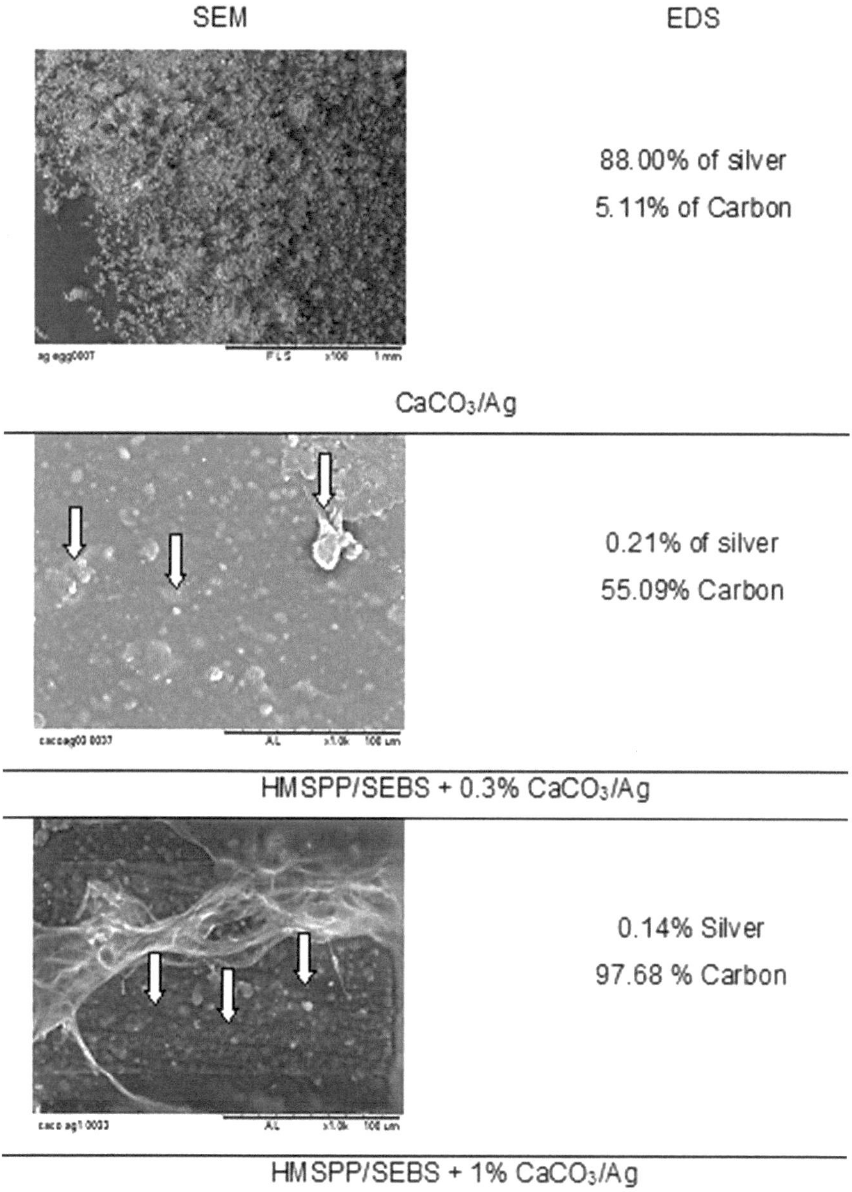

Fig. 2 SEM and EDS of $CaCO_3$/AgNp powder and surfaces of the injected molded samples with $CaCO_3$/AgNp

Table 2 DSC and TGA analysis of samples

Samples	T_c/°C	T_{m2}/°C	X_c/%	T_{onset}/°C 1°/2°
HMSPP/SEBS	110.86	155.31	37.39	213.60/384.30
HMSPP/SEBS+0.3% $CaCO_3$	111.38	156.15	30.36	213.08/398.83
HMSPP/SEBS+1% $CaCO_3$	111.66	154.37	28.13	209.12/395.55
HMSPP/SEBS+0.3% $CaCO_3$/Ag	111.38	156.15	30.36	220.54/405.98
HMSPP/SEBS+1% $CaCO_3$/Ag	112.23	155.99	34.97	175.38/408.11

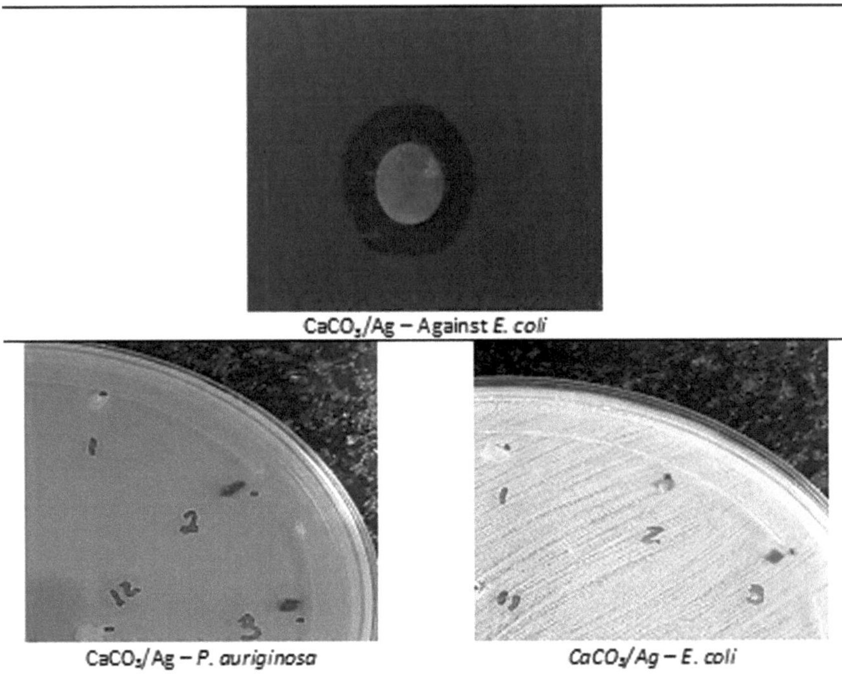

Fig. 3 Microbiological results on injection molded samples and on the powder $CaCO_3$/AgNp

The key point to observe related to the microbiological activity in the presence of the $CaCO_3$/Ag and $CaCO_3$ before and after the processing can be seen in Fig. 3.

The formation of inhibition zone was observed in the microbiological test of the particle alone. However, the same did not occur on the injection-molded particles in polymer composition. This can be attributed to the fact that $CaCO_3$/Ag is agglomerated and coated by the polymer.

Conclusion

The results indicated that the fine particles have a microbiological effect, even if the size is micrometer scale, as observed on DLS analysis. It was interesting to observe that processing affects drastically the Ag activity.

References

1. Narayanan KB, Suresh AK, Sakthivel N (2015) Metallic nanocomposites: bacterial-based ecologically benign biofabrication and optimization studies. In: Eco-friendly polymer nanocomposites. Advanced Structure Materials. Springer, India, pp 215–231
2. Tiimob BJ, Jeelani S, Rangari VK (2016) Eggshell reinforced biocomposite—an advanced "green" alternative structural material. J Appl Polym Sci. https://doi.org/10.1002/app.43124
3. Nomiya K, Kasuga NC, Takayama A (2014) Synthesis, structure and antimicrobial activities of polymeric and nonpolymeric silver and other metel complexes. In: Polymeric materials with antimicrobial activity: form synthesis to applications. RSC Polymer Chemistry Series, pp 156–207
4. Guzmán MG, Dille J, Godet S (2009) Synthesis of silver nanoparticles by chemical reduction method and their antibacterial activity. Int J Chem Bio Eng, 104–111
5. Le AT, Tam LT, Tam PD, Huy PT, Huy TQ, Hieu NV, Cotolan N, Rak M, Bele M, Cör A, Muresan LM, Milosev I (2016) Sol-gel synthesis, characterization and properties of TiO_2 and Ag-TiO_2 coatings on titanium substrate. Surf Coat Technol, 790–799
6. Kudrinskiy AA, Krutyakov YA (2010) Synthesis of oleic acid-stabilized silver nanoparticles and analysis of their antibacterial activity. Mater Sci Eng C, 910–916

The Potential of Micro- and Nano-sized Fillers Extracted from Agroindustry Residues as Reinforcements of Thermoplastic-Based Biocomposites—A Review

Esperidiana A. B. Moura

Abstract Currently, the relevance of reuse of agroindustrial waste to obtain fillers in micro- and nano-sizes for the development of biocomposite materials has grown significantly. Production processes based on sustainable and low carbon development have increased interest in more environmentally friendly polymer composites, which have made the origin of reinforcement materials a determining factor for their application in this segment. This work presents a review of the developments of our team in the field of thermoplastic biocomposites reinforced with micro- and nano-sized fillers extracted from agroindustry residues. The different residues from Brazilian agroindustry available for the extraction of micro- and nano-sized fillers for the production of polymer biocomposites, the methods of the extraction and treatments of these natural fillers are presented; and its application as reinforcements in thermoplastic-based biocomposite are discussed in this review.

Keywords Agroindustry waste · Biocomposites · Composites · Natural fillers, thermoplastic-based biocomposites

Introduction

Currently, more environmentally friendly polymer composites that use raw materials from renewable sources that do not harm the environment have driven research and development of new materials. Interest in these materials is due to increased attention to production processes characterized by sustainable, low carbon development, because of concerns about air, water, soil and food quality, and the reduction of pollutants to maintain the environment with the least quality for future generations. A complicating factor of this problem is that over the years, the high consumption of plastic products produced from conventional non-biodegradable polymers has produced a large residual mass that accumulates in landfills, rivers, and seas, due to inappropriate

---

E. A. B. Moura (✉)
Center for Chemical and Environmental Technology, Nuclear and Energy Research Institute, 2242 Prof. Lineu Prestes Av, São Paulo 05508-000, Brazil
e-mail: esperimoura@gmail.com

S. Ikhmayies et al. (eds.), *Green Materials Engineering*, The Minerals, Metals & Materials Series, https://doi.org/10.1007/978-3-030-10383-5_10

disposal, generating problems environmental impacts. The non-biodegradability of materials produced from conventional polymers further aggravates these problems, since it may take years for complete decomposition [1, 2].

In this context, the origin of reinforcement materials is becoming determining factors for their application, and important demands are expected for those applications where biodegradability and/or more friendly composites offer clear advantages to customers and the environment, for which recovery and recycling are a problem. The global attention to the end-of-life of any product on the market, and the application of the "3 Rs" of sustainability (Reduce, Reuse, and Recycle), has further boosted the research and development of greener composites.

Currently, polymer matrix composites, obtained by the dispersion of micro- and/or nano-sized fillers, extracted from residues of mineral activity or agribusiness are a category of materials of great interest, even for durable applications that require more rigorous technical performances, particularly in terms of mechanical properties and stability [2, 3].

In this presents review, the developments of our team in the field of thermoplastic biocomposites reinforced with micro- and nano-sized fillers extracted from agroindustry residues were summarized. The different residues from Brazilian agroindustry available for the extraction of micro- and nano-sized fillers for the production of polymer biocomposites, the methods of the extraction, and treatments of these natural fillers were presented; its applications as reinforcements in thermoplastic-based biocomposites were also discussed [4–12].

Thermal and Morphological Behavior of PBT/Sugarcane Bagasse Ash Composites

The incorporation of sugar bagasse ashes obtained from the burning process of sugarcane bagasse to produce energy in the sugar and bioethanol industry in the poly(butylene terephthalate) PBT resulted in no important visual changes on the surface of the cryofractured PBT. Some decrease in polymer crystallinity was observed by DSC analysis, but the most important result was the increase in the heat deflection temperature of over 36%. Such increase is of great practical interest once most of the PBT used today is reinforced with fiber glass to guarantee its thermo-mechanical performance (Table 1).

The addition of the ashes in this study showed that it is possible to get interesting property gains by using waste from renewable sources instead of the traditional ones. Changes in the polymer thermal and morphological behaviors can be related to the high content of silica (SiO_2) in the ashes which corresponds to 57% of the ash composition in the studied waste [5].

Table 1 Mechanical and thermo-mechanical results for the neat PBT and PBT/ash composite. (Adapted from Ref. [5])

Test	Neat PBT	PBT/Ash composites	Variation (%)
HDT (°C)	55.4	75.7	+36.6
Tensile strength at break (MPa)	38.0	56.1	+47.6
Elongation at break (%)	161.6	16.2	−90.0
Flexural strength (MPa)	74.2	86.4	+16.4
Flexural modulus (GPa)	2.4	2.9	+20.8

Short Vegetal-Fiber Reinforced HDPE Treated by Electron-Beam Radiation

Piassava (Attalea Funifera Mart) is a Brazilian lignocellulosic fiber extracted from the leaves of a palm tree of natural occurrence in the Atlantic rain forest, and its exploitation is an extractive activity that represents the main source of income to approximately 2000 small-scale farmers, processors, and their families. Piassava fibers have been described as harder than other lignocellulosic fibers and have higher lignin content (around 48%) than any of the other common lignocellulosic fibers. In this paper, short piassava fibers (30 wt%) were incorporated into high-density polyethylene (HDPE) by melt extrusion using a twin-screw extruder extrusion processing, and glycidyl methacrylate (GMA) was added at 2.5 and 5.0% (on piassava fiber wt.) as a cross-linking agent, and their effects upon the properties of the resulting composites treated by electron-beam radiation were also examined [6]. These results showed an increase of around 250% in tensile strength at break of neat HDPE due to the addition of the piassava fiber with GMA, and of around 300% for the composites consisting of HDPE reinforced with piassava fiber and GMA treated by electron-beam irradiation. It can be distinctly observed that the addition of piassava fiber with GMA led to the increase in Young's modulus of around 100% and of around 130%, when the composites are also treated by electron-beam irradiation (Fig. 1). Examination of the composite failure surfaces indicated that there was an improved adhesion between fiber and matrix with the radiation dose and GMA content. It is clear from these results that, as widely reported in the literature, electron-beam irradiation induced predominantly cross-linking in the amorphous regions of HDPE, whereas the cross-linking effectively increases intermolecular bonds in this region, resulting in an improvement in material properties, such as better mechanical and morphological properties [2, 6].

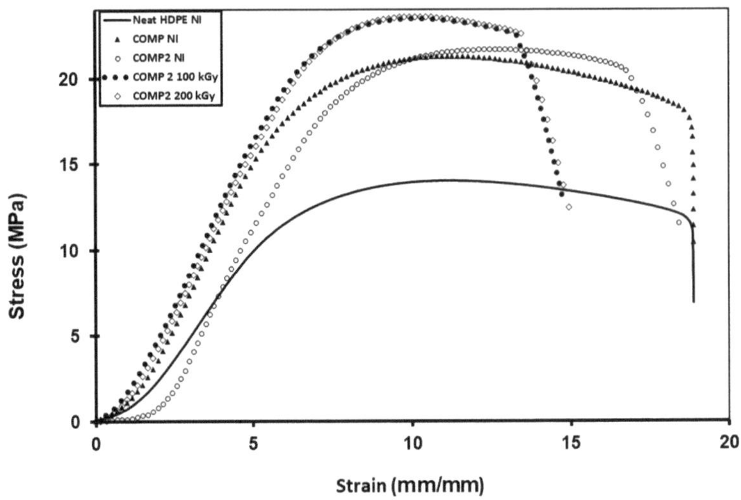

Fig. 1 Diagram Stress (MPa) X Strain (mm/mm) for irradiated and non-irradiated HDPE/Piassava fiber composites: COMP NI = HDPE/piassava fiber; COMP 2 = HDPE/5.0GMA/piassava fiber. (Adapted from Ref. [6])

Copolyester/Starch Blend Reinforced with Rice Husk Ash

The synthesis of potentially biodegradable polymers is of considerable significance and has been receiving much attention. However, the mechanical properties of biodegradable polymers are very poor for many applications. Biocomposites obtained by the combination of biodegradable polymer as the matrix material and biodegradable fillers (rice husks, bagasse ashes) as the integral part are also expected to be biodegradable, since both components are biodegradable and can contribute to reduce the use of non-biodegradable plastic materials and improving management of solid waste disposal. Rice husk is an important by-product of rice milling process and is the major waste product of the agricultural industry. Rice husk contains nearly 20 wt% of silica, which is present in hydrated amorphous form. Currently, rice husk and sugarcane bagasse are widely used in the energy co-generation process, and as a result of its burning, tons of ashes are produced. These ashes can be used directly as reinforcement of polymer or processed for obtaining "green silica", in the form of micro- and nanoparticles for addition in polymeric materials [7].

In this study, rice husk ashes were added in copolyester/starch blend by melt extrusion using a twin-screw extruder extrusion processing in order to improve the mechanical and morphological behaviors of this biodegradable blend [7]. The results showed a gain of crystallinity of the polymeric blend, changes in the morphological behaviors, and a significant gain in tensile strength at break and Young's modulus (Table 2). According to the results of this study can be inferred that rice husk ashes, a renewable agro-resource of low cost, a widely available raw material from energy

Table 2 Mechanical test results of PBAT/starch blend and PBAT/starch/ash composite. Adapted from Ref. [7]

Properties	Neat copolyester/starch blend	Composite
Tensile strength at break (MPa)	6.64 ± 0.44	9.96 ± 0.7
Elongation at break (%)	83.22 ± 12	11.12 ± 0.5
Yong modulus (MPa)	35.35 ± 0.2	99.9 ± 2.7
Notched Izod impact (J/m)	*	35.62 ± 2.5

*PBAT/starch blend is a non-breakable extremely flexible material

co-generation processes, are a good candidate for the reinforcement of biodegradable plastics in order to obtain composite materials with advanced mechanical behaviors for several applications.

Nanocellulose Extracted from Sugarcane Bagasse and Its Incorporation into Polymers

Nanocellulose extracted from agroindustry residues has attracted attention of the industry and the academy since its application contributes to the development of sustainable nanocomposite materials, while promoting reduction of the environmental impact caused by non-utilization of these residues. However, nanocellulose as nanofiller usually tends to form agglomerates due to the attractive Van der Waals forces, as intra- and intermolecular hydrogen bonds and hydrophilic nature that reduce its dispersion and distribution in polymer matrix. The surface treatment and chemical functionalization can be used as measures to increase hydrophobicity and minimize the above-mentioned interactions.

Recently, our group have extracted nanocellulose from sugarcane bagasse an agro-waste from Brazilian alcohol and sugarcane industry, functionalized by grafting of glycidyl methacrylate (GMA) polymerization induced by gamma radiation. Natural nanocellulose extracted from sugarcane bagasse and grafted with GMA by gamma radiation was incorporated by melt extrusion processing into biodegradable flexible films prepared from poly (butylene adipate co-terephthalate) PBAT/poly (lactic acid) PLA blend, and also incorporated into conventional flexible films based on EVA and LDPE [8, 9]. The results demonstrated that nanocellulose extracted from sugarcane bagasse can be used to prepare biodegradable flexible films with enhanced mechanical properties for food and cosmetic packaging application (Fig. 2).

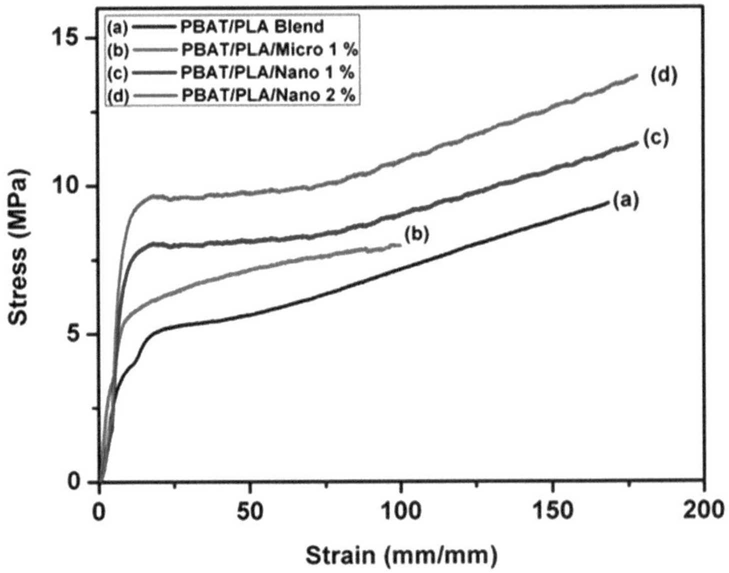

Fig. 2 Diagram stress (MPa) X strain (mm/mm) for PBAT/PLA blend and its composites. (Adapted from Ref. [8])

Nanocellulose grafted with GMA by gamma radiation when incorporated into biodegradable and conventional polymer improved the bonding between the nanofillers and matrix and, consequently, led to the production of flexible films with more desirable properties for application on food and cosmetic packaging (Figs. 3 and 4).

Composite Materials Development of Polymer Foams Reinforced with Babassu Coconut Fiber Residues

Composites based on recycled-HDPE/EVA blend reinforced with babassu coconut fiber were obtained by melting extrusion process. The composites were then extruded in a special single screw for foaming. The foam samples were subjected to mechanical tests, density measurement, DSC, TG, and FE-SEM analysis. The results showed an increase of density of foam and a closed-cell structure with a relatively homogeneous cell size distribution due to babassu fiber residues addition. The foams prepared from HDPEr/EVAr blend presented a mixed morphology of open-cell and closed-cell structures, but with superior tensile strength and elongation properties when compared with foam from HDPEr/EVAr/Babassu composites [10].

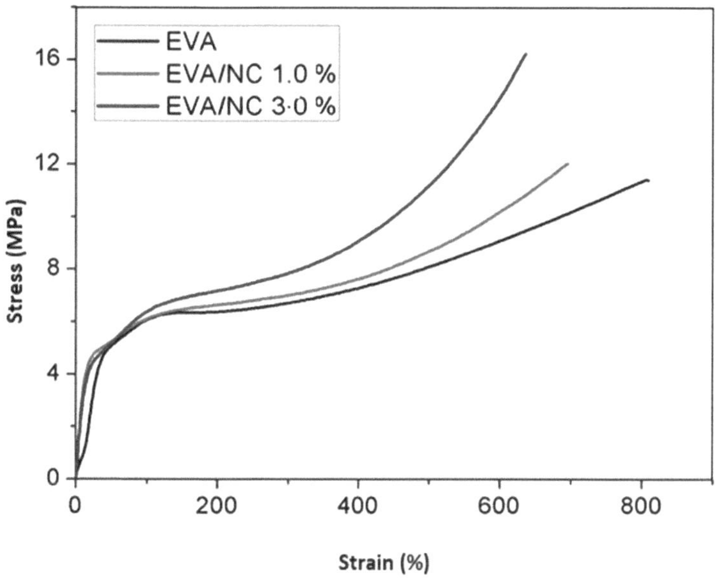

Fig. 3 Diagram stress (MPa) X strain (%) for neat EVA and EVA/Nanocellulose flexible films. (Adapted from Ref. [9])

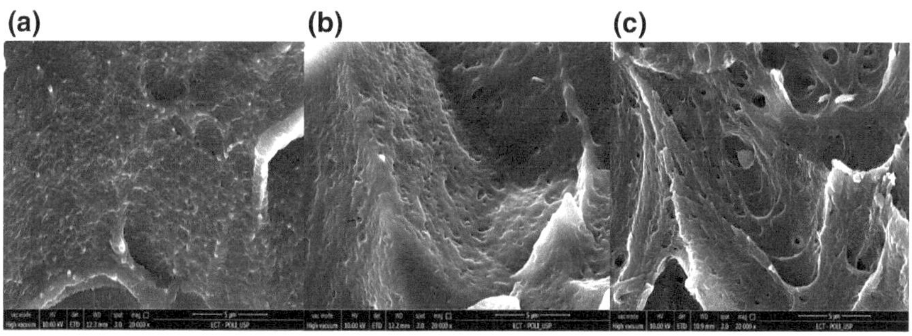

Fig. 4 FE-SEM images for neat EVA (a) and EVA/Nanocellulose (1 wt.%), EVA/Nanocellulose (3 wt.%) flexible films [9]

Recycled Polymer Reinforced with Natural Fibers from Agricultural Wastes

Agricultural residues, which are produced with large quantities annually throughout the world, may be used as reinforcement plastic to replace the wood and produce particle board for application in the development of low-cost construction elements and reduced environmental impact. The main aims of this study were to investigate the effects of agricultural wastes and glass residues addition to the properties of recycled-

HDPE/EVA blend for use in particle board manufacture. The recycled-HDPE/EVA blend reinforced with corncob fiber (15 wt%), coffee parchment (15 wt%), and glass residues (1 wt%) was processed by melt extrusion, using a twin-screw extruder and injection molding machine to obtain specimen test samples. The results showed gains in tensile strength at break, elongation at break, and Young's modulus properties of blend due to agricultural residues addition. Compared with the r-HDPE/EVA blend, the endothermic melting enthalpy of its composite decreased considerably due to agricultural residues addition. The decreases in the melting enthalpy can be attributed to the increase in crystallinity of r-HDPE/EVA composite. XRD spectrum of r-HDPE/EVA blend presented a large reduction of the intensity due to agricultural and glass residues addition. FE-SEM showed homogeneous distribution of agricultural fiber residues particles, but with some holes in r-HDPE/EVA blend due to agricultural fiber residues pull out [10].

Eggshells a Potential Source of Bio-fillers Highly Valuable Polymeric

Eggshell is an important waste material by-product of poultry industries and domestic kitchen waste. Currently, egg production throughout the world is 65.5 million metric tons per year, with Asia as a key contributor to global egg output growth. By taking 11% of the weight, nearly 7.2 million tons of eggshell waste is created every year; these residue materials, if unutilized, create a potential pollution problem. Therefore, resourceful utilization of this waste is of great importance not only for reducing the environmental impact, but also for gaining higher profits. Eggshell is a natural bioceramic composite with a unique chemical composition of high inorganic (95% of calcium carbonate in the form of calcite) and 5% of organic (type X collagen, sulfated polysaccharides) components; this eggshell characteristic structure combined with substantial availability makes eggshells a potential source of bio-fillers that can be efficiently used for polymer composites [11, 12].

Bio-calcium Carbonate from Eggshells as Reinforcement of Biodegradable Polymer

This study aims to the development of bio-foams from PBAT/PLA blends reinforced with bio-calcium carbonate from eggshells [11]. The addition of calcium carbonate from avian eggshells proved to be an effective reinforcement for PBAT/PLA matrix, according to tensile and compression tests. MFI assessments confirmed a lower value for PBAT/PLA matrix, after insertion of calcium carbonate, indicating a lower molecular weight distribution and consequently a higher molar mass. A very higher crystallinity (from around 5 to 33%) was observed after blending PBAT and PLA (Table 3). Thermal behavior was ineffective to indicate a maximum degradation

Table 3 Thermal and crystallinity properties for PBAT/PLA and their composites. (Adapted from Ref. [11])

Properties	PLA	PBAT	PLA*	PBAT/PLA	PBAT/PLA + $CaCO_3$	PBAT/PLA + $CaCO_3$ + PLA*	PBAT/PLA + $CaCO_3$ + PLA*
				(50/50) (wt%)	(85 + 15) (wt%)	(80 + 15 + 5) (wt%)	(80 + 15 + 10) (wt%)
Tcc (°C)	33.5	51.2	39.8	71.7	69.7	69.8	70.0
ΔHc ($J.g^{-1}$)	36.8	20.5	43.5	7.2	7.1	7.3	7.5
Tm (°C)	87.8	120.0	86.9	86.9	87.4	87.8	87.2
ΔHm ($j.g^{-1}$)	43.5	16.8	41.6	22.5	21.2	17.2	10.7
X (%)	7.2	3.2	2.1	33.0	33.0	33.0	33.0

PLA* = irradiated PLA; Tcc = cold crystallization temperature; Tm = melt temperature
ΔHm = melt enthalpy; X = crystallinity

temperature due to very close results. DRX confirmed previous DSC crystallinity results for PBAT/PLA and composites.

Biodegradable Polymer Reinforced with Bio-hydroxyapatite from Eggshell Waste

The calcium carbonate from avian eggshells may be used in bone regeneration in the natural form or converted to calcium phosphates, classified as bio-hydroxyapatite for bone healing in dentistry and orthopedic field. The bone tissues in vertebrates are composed mainly of bio-hydroxyapatite, which explains why these calcium phosphates materials have chemical properties suitable for bone regeneration. A bio-hydroxyapatite (HAp) is one of the main inorganic ceramics that make up this system, having its origin from calcium phosphate with the chemical structure (PO_4) $6(OH)_2$. Its application as a bone regenerator develops from its construction, which is responsible for about 7% of the constitution of the bones and teeth of up to 90% of the enamel of the teeth, when found naturally in the human body [12]. As a result, this apatite has become the main biomaterial used for reconstruction of damaged tissues or when there is a loss of bone mass. Thus, a HAp is an ideal candidate for orthopedic implants, dental procedures, surgical procedures, grafts, and biomarkers applications. Nowadays, significant advances have been made in the development of biodegradable polymeric materials for biomedical applications.

This study aims to prepare and characterize composite materials based on PLA/PBAT, a biodegradable polymer blend, reinforced with bio-hydroxyapatite (bio-HAp). Firstly, bio-HAp was obtained from eggshell residues. PLA/PBAT blend with 1–5 wt% of bio-HAp was prepared by melt extrusion and injection molding process.

Table 4 Tensile tests results for the PLA/PBAT and their composites. (Adapted from Ref. [12])

Materials	Tensile strength at break (MPa)	Young's modulus (MPa)	Elongation at break (%)
PLA/PBAT BLEND	17.8 ±0.4	582.8 ±1.1	40.2 ±3.5
PLA/PBAT/HAp 1%	18.3 ±0.3	599.2 ±0.9	41.3 ±2.7
PLA/PBAT/HAp 3%	19.5 ±0.2	638.5 ±0.8	44.1 ±3.4
PLA/PBAT/HAp 5%	20.4 ±0.2	667.9 ±1.0	46.2 ±2.8

The effects of the bio-HAp addition into the biodegradable blend were investigated by mechanical tests, XRD, DSC, FE-SEM, and cytotoxicity "in vitro" analysis, and the correlation between the properties was discussed. The composites showed an increase in tensile properties up to 15 wt% when compared with PLA/PBAT blend (Table 4); a rough, dense and compact micrographs surface with HAp particles dispersed in the matrix and with only some small agglomerates of HAp particles, indicating a good dispersion of bio-HAp into PLA/PBAT matrix [12].

Based on results, it can be concluded that the PLA/PBAT reinforced with bio-HAp produced from eggshell wastes presents improved mechanical properties, good dispersion of HAp particles into the matrix and potential to be used as biomaterials. In Fig. 5, it can be seen any significant difference between PLA/PBAT and their composites, thus presenting cell viability greater than 75%, allowing them to be classified as non-cytotoxic according to 23rd US pharmacopeia. Then, it can be concluded that PLA/PBAT and their composites have the potential to be used as biomaterials [12].

Conclusions and Outlook

This review presented the developments of our team in the field of thermoplastic biocomposites reinforced with micro- and nano-sized fillers extracted from agroindustry residues. The different residues from Brazilian agroindustry available for the extraction of micro- and nano-sized fillers for the production of polymer biocomposites, the methods of the extraction and treatments of these natural fillers were shown, and their application as reinforcements in thermoplastic-based biocomposite was also discussed in this review. The results from these studies showed the viability of the reuse of agroindustry residues as raw materials for the development of environmentally friendly composite materials; while avoids the criminal discards and the proliferation of diseases caused by bacteria and fungi. Their uses can lead to the development of ecologically correct plastic materials that add value and meet the expectations and needs of society. The composites obtained have a range of applications, from the food and cosmetic packaging segment up to biomedical applications such as the manufacture of biomaterials with excellent properties for the regeneration of bone and dental tissues, support for drug delivery and additives for the

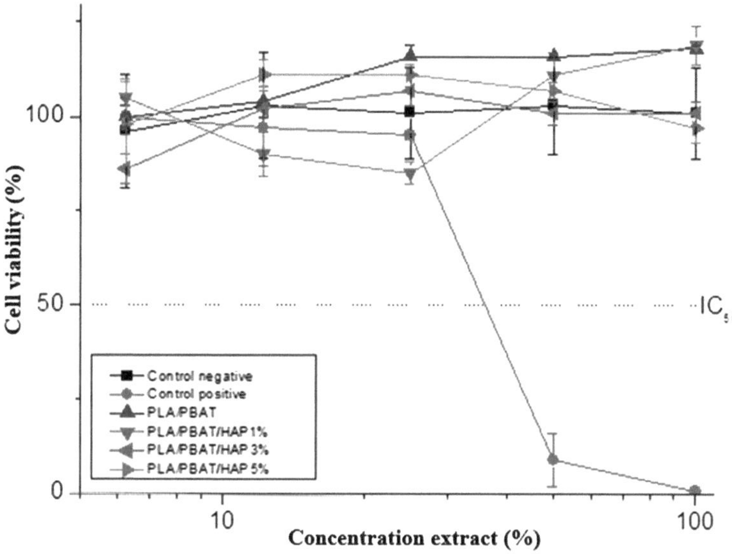

Fig. 5 Cell viability curves obtained in the cytotoxicity assay by the neutral red incorporation method. (Adapted from reference [12])

manufacture of dental creams. The replacement of conventional composite materials by these environmentally friendlier ones presents yet the additional advantages of being renewable, recyclable, and of great environmental and social importance.

Acknowledgements The authors wish to thank IAEA-CRP# 17760 RO, CAPES, and CNPq for the support for this work.

References

1. Ortiz AV, Teixeira JG, Gomes MG, Oliveira RR, Valenzuela-Díaz FR, Moura EAB (2014) Preparation and characterization of electron-beam treated HDPE composites reinforced with rice husk ash and Brazilian clay. Appl Surf Sci 310:331–335
2. Güven O, Monteiro SN, Moura EAB, Drelich JW (2016) Re-emerging field of lignocellulosic fiber – polymer composites and ionizing radiation technology in their formulation. Polym Rev (6)1:1560
3. Kotal M, Bhowmick AK (2015) Polymer nanocomposites from modified clays: recent advances and challenges. Prog Polym Sci 51:127–187
4. Campos RD, Sotenko M, Hosur M, Jeelani S, Valenzuela Díaz FR, Moura EAB, Kirwan K, Seo ESM (2015) Effect of mercerization and electron-beam irradiation on mechanical properties of high density polyethylene (HDPE)/Brazil nut pod fiber (BNPF) biocomposites. In: Carpenter JS et al (eds) Characterization of minerals, metals, and materials 2015. The minerals, metals & materials society. Wiley, Hoboken, pp 637–644

5. Ortiz AV, Pozenato CA, Sartori MN, Oliveira RR, Scarpin MA, Moura EAB (2012) Thermal and morphological behavior of PBT/sugarcane bagasse ash composite. J Nanostructured Polym Nanocomposites 8:78–81
6. Ferreira MS, Sartori MN, Oliveira RR, Guven Olgun, Moura EAB (2014) Short vegetal-fiber reinforced HDPE—a study of electron-beam radiation treatment effects on mechanical and morphological properties. Appl Surf Sci 310:325–330
7. Oliveira EH, Silva VA, Oliveira RR, Soria A, Abreu A, Harada J, Diaz FRV, Moura EAB (2014) Investigation on mechanical and morphological behaviours of copolyester/starch blend reinforced with rice husk ash. In: Carpenter JS et al (eds) Characterization of minerals, metals, and materials 2014. The minerals, metals & materials society. Wiley, Hoboken, pp 491–498
8. Paiva DA, Oliveira RR, Silva WM, Auad ML, Rangari Vijaya K, Moura EAB (2016) Comparative study of the effects of cellulose nanowhiskers and microcrystalline cellulose addition as reinforcement in flexible films based on biopolymer blends. In: Ikhmayies SJ et al (eds) Characterization of minerals, metals, and materials 2016. The minerals, metals & materials society. Wiley, pp 409–416
9. Seixas MVS (2018) Obtainment of Nanocellulose from Sugarcane Bagasse and its incorporation into EVA. PhD thesis. Escola Politécnica da Universidade de São Paulo. Departamento de Engenharia Metalúrgica e de Materiais
10. Arantes M, Santana JG, Valenzuela-Díaz F, Rangari VK, Guven O, Moura EAB (2018) Development and characterization of recycled-HDPE/EVA foam reinforced with babassu coconut epicarp fiber residues. In: Li B et al (eds) Characterization of minerals, metals, and materials 2018, The minerals, metals & materials series. Springer International Publishing AG, Switzerland, pp 497–506
11. Cardoso ECL, Seixas MVS, Wiebeck H, Oliveira RR, Machado GAF, Moura EAB (2016) Development of bio-based foams prepared from PBAT/PLA reinforced with bio-calcium carbonate compatibilized by electron-beam radiation. In: Ikhmayies SJ et al (eds) Characterization of minerals, metals, and materials 2016. The minerals, metals & materials society. Wiley, pp 637–644
12. Reis PRS, Santana JG, Oliveira RR, Rangari VK, Lourenço FR, Moura EAB (2019) Development of biocomposite materials from biodegradable polymer and bio-hydroxyapatite derived from eggshells for biomedical applications. Submitted to publish in: The minerals, metals & materials series, 1st edn. Springer International Publishing, Switzerland, February 2019

Thermal Behavior of Epoxy Composites Reinforced with Fique Fabric by DSC

Michelle Souza Oliveira, Artur Camposo Pereira, Sergio Neves Monteiro, Fabio da Costa Garcia Filho and Luana Cristyne da Cruz Demosthenes

Abstract In recent years, there has been a remarkable growth in the development of composite materials reinforced by natural fibers, especially by the substitution of synthetic fiber such as glass and carbon fiber, which are commonly used reinforcement materials for composites. Despite the benefits associated with the use of natural fibers, there are still some limitations to their application. Among the disadvantages presented by natural fibers stands the low thermal resistance. The hydrophilic characteristic of these natural lignocellulosic fibers (NLF) causes absorption of water, but at high temperatures, this water is lost and tends to produce pores and flaws in the composite. The objective of this work is to study the thermal behavior of epoxy composite materials reinforced with NLF in the proportions of 15, 30, 40, and 50 vol%. The samples were characterized by differential scanning calorimetry (DSC). The DSC curves suggest that the variation of the amount of natural fibers used for reinforcement did not affect the thermal stability of the composite once any variation of enthalpy was observed.

Keywords DSC analysis · Fique fabric · Composites · Epoxy matrix

Introduction

Nowadays polymer matrix composite materials reinforced with fibers such as glass or carbon present specific problems, their high dependence on oil and the waste management after its end of life. Thereby, natural fibers are becoming matters of great interest due to their wide range of applications [1–10].

Some characteristics that make natural fibers attractive are low cost, their high specific stiffness and impact resistance, reduced energy consumption, non-toxicity, renewability, recyclability, and biodegradability. They also provide thermal and

M. S. Oliveira (✉) · A. C. Pereira · S. N. Monteiro · F. da Costa Garcia Filho · L. C. da Cruz Demosthenes
Military Institute of Engineering–IME, Rio de Janeiro, Brazil
e-mail: oliveirasmichelle@gmail.com

S. Ikhmayies et al. (eds.), *Green Materials Engineering*, The Minerals, Metals & Materials Series, https://doi.org/10.1007/978-3-030-10383-5_11

acoustic insulation, structural lightness, which results in a reduction in weight and therefore, lower fuel consumption and lower emissions for specific applications [4].

Natural and synthetic fibers when subject to higher temperatures tend to suffer degradation, which has a direct impact on the mechanical properties of the material. Lignocellulosic fibers will begin seeing significant low degradation temperatures around 180–210 °C [11], which make them incompatible with thermosets that have high curing temperatures. The most thermosets matrices used for natural fiber composite materials are unsaturated polyester, epoxy resin, phenol formaldehyde, and vinyl ester resins [10].

Among the lignocellulosic fibers with potential for composite reinforcement that extracted from South American Andean region, fique is one of the most abundant fibers and its main uses are in the manufacture of sacks, shoes, bags, and handicrafts [12].

The first studies carried out on fique fiber are about the influence of surface cationization conditions [3], chemical treatments (mercerization, maleic anhydride, acrylic acid, and silane) on the mechanical, chemical, and physical properties of fique fibers, and also on the wettability of the untreated and treated fique fibers [7, 13, 14].

The influence of fiber length, content, and surface treatments on mechanical and physical behavior of fique fibers reinforced polypropylene matrix composites have also been studied [8]. Moreover, alkaline treatment under tension for fique fibers and its effect on mechanical properties were analyzed from changes in structure and morphology of the fibers [15, 16] and the use of dry etching plasma for treating fique fibers, in which the exposure time to the same energy was variable [17].

Furthermore, the fractographic analysis of the fique fiber and composite material developed with fique fabrics as reinforcement and with polyester resin as matrix were developed [4], failure analysis of machinery supports for fique processing [5], and experience in applying an appropriate low scale production technology for manufacturing corrugated sheets for roofing using vacuum [6].

As to fiber/matrix adhesion was reported an improvement by modification of the fiber surface properties utilizing chemical treatments with polyethylene prior to preparation of the composite materials. The effect of the fiber/matrix interface in the creep behavior and the dynamic chemical properties were investigated for a composite material based on a matrix of LDPE–Al reinforced with fique fibers [18].

In addition to brief review mentioned, the literature presents the mechanical properties of polyester matrix composites reinforced with fique fabric by means of Izod impact tests [19] and Charpy test [12] as well as the dynamic behavior, considering the application as a ballistic armor [20]. On the other hand, the incorporation of nonwoven industrial fique fiber mats and their effects on the thermal and mechanical properties of two polymer matrices with a different thermal behavior (LLDPE and a flexible EP) also has been reported [10]. Furthermore, mechanical properties regarding epoxy matrix composites reinforced with continuous and aligned fique fibers in terms of Charpy impact test [1], bending test [2], and the fique fiber tensile elastic modulus dependence with corresponding diameter can be found in the literature [21].

Aforementioned many studies have been carried out regarding the properties of fique fiber, especially when used as reinforcement to polymer matrix composites. Yet some properties about this material are still in lack of investigation. In this context, the purpose of this work was to study the influence of the incorporation of fique fabric on the thermal properties of polymer matrix composites. This would allow appreciating the real possibilities for using this fiber as reinforcement.

Materials and Methods

Materials Source and Process

Fique fabric was obtained from the local market, in Colombia. The fabric was dried at 60 °C in an air-oven for 24 h in order to reduce the moisture content. This process favors the adhesion between the matrix and the natural fiber.

Composite Preparation

Composites with different volume fractions (15, 30, 40, and 50 vol%) of fique fabric as reinforcement were produced using a compression molding process. The epoxy resin was poured inside the mold and allowed to cure at room temperature for 24 h under a 3 MPa tension. After extraction from the mold, the composite was sectioned for each fiber percentage for the epoxy matrix.

Differential Scanning Calorimetry (DSC)

DSC test was carried out using Netzsch DSC 204 F1 Phoenix equipment differential scanning calorimeter under nitrogen atmosphere operating from 80 to 200 °C at a scanning rate of 10 °C/min, with a sample of 10 mg in aluminum pans. This experiment was performed to determine the extent of cure versus cure time curves for the different composites.

Results and Discussion

Figure 1 shows the set of first run DSC curves for the fique fiber and the epoxy matrix composites reinforced with 15, 30, 40, and 50 vol% of fique fabric. This figure displays the DSC curve for all conditions investigated; one may notice the existence of a slightly endothermic peak around 147 °C for fique fiber. For the

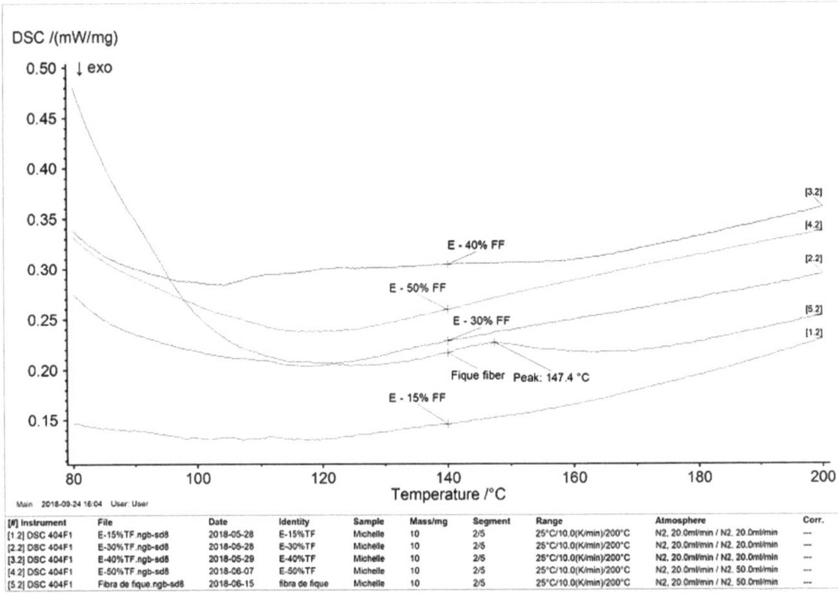

Fig. 1 Differential scanning calorimetry (DSC) curves of fique fiber and its composite

composites analyses, it was not observed any relevant event for this analysis within the stipulated temperature range.

Ovalle-Serrano et al. [22] reported the following events in the DSC analysis in fique fibers:

Temperature (°C)	Peak	Process
30–130	Endothermic	Water evaporation
165–260	Exothermic	Hemicellulose degradation
360	Endothermic	Fusion and degradation of crystalline cellulose

Fibers showed a higher final temperature (90 °C) for water evaporation perhaps due to the presence of amorphous components (hemicellulose and lignin) that help to retain moisture by chemi- and physisorption processes. Hemicellulose degradation occurs in fique fibers related to α-cellulose depolymerization (decomposition) and formation of 1,6-anidroglucose.

Hidalgo-Salazar et al. [10] reported that neat epoxy and epoxy/fique composite do not crystallize on cooling. The glass transition temperature (Tg) of the epoxy at about 31 °C while for the composite epoxy/fique this temperature is around 48 °C. The glass transition temperature of a polymer is known to depend on the mobility of the chain segment of the macromolecules in the polymer matrix. In this case, the reinforcement effect of the fique fibers restricts the motion of epoxy macromolecule chains and thus increases the Tg of the composite.

Summary and Conclusions

- DSC curves of fique fabric reinforced epoxy composites were found to be similar for all tested conditions.
- The curves did not present any relevant event for this analysis within the stipulated temperature range, which suggests that the glass transition temperature of the composite was displaced due to the presence of the fique fabric reinforcement.

Acknowledgements The authors thank the Brazilian agencies CAPES and CNPq for the financial support.

References

1. Altoé GR, Teles MC, Ferreira MV, Fernandes GV, Margem FM, Monteiro SN (2016) Charpy impact tests in epoxy matrix composites reinforced with continuous fique fiber. 71° Congresso Anual da ABM—Internacional e ao 16° ENEMET—Encontro Nacional de Estudantes de Engenharia Metalúrgica, de Materiais e de Minas, 839–845
2. Altoé GR, Teles MC, Netto PA, Margem FM, Monteiro SN (2015) Bending tests in epoxy composites reinforced with fique fibers. 70° Congresso Anual da ABM—Internacional e ao 15° ENEMET—Encontro Nacional de Estudantes de Engenharia Metalúrgica, de Materiais e de Minas, 742–748
3. Castellanos LJ, Blanco-Tirado C, Hinestroza JP, Combariza, MY (2012) In situ synthesis of gold nanoparticles using fique natural fibers as template. Cellulose, 1933–1943
4. Contreras MF, Hormaza WA, Marañón A (2009) Fractografía De La Fibra Natural Extraida Del Fique De Un Material Compuesto Reforzado Con Tejido De Fibra De Fique Y Matriz Resina Poliester. Suplemento De La Revista Latinoamericana De Metalurgia Y Materiales, 57–67
5. Delgado F, Coronado J, Rodríguez S (2015) Failure analysis of a machine support for fique fibre processing. Eng Fail Anal. https://doi.org/10.1016/j.engfailanal.2015.04.012
6. Delvasto S, Perdomo EF, Guiérrez RM (2010) An appropriate vacuum technology for manufacture of corrugated fique fiber reinforced cementitious sheets. Constr Build Mater 24:187–192
7. Gañán P, Mondragon I (2002) Surface modification of fique fibers. Effects on their physico-mechanical properties. Polym Compos, 383–394
8. Gañán P, Mondragon I (2004) Influence of compatibilization treatments on the mechanical properties of fique fiber reinforced polypropylene composites. Int J Polym Mater Polym Biomater 53(11):997–1013. https://doi.org/10.1080/00914030490516648
9. Hidalgo-Salazar MA, Correa JP (2008) Mechanical and thermal properties of biocomposites from nonwoven industrial fique fiber mats with epoxy resin and linear low density polyethylene. Results Phys 8:461–467. https://doi.org/10.1016/j.rinp.2017.12.025
10. Hidalgo-Salazar MA, Correa JP (2018) Mechanical and thermal properties of biocomposites from nonwoven industrial fique fiber mats with epoxy resin and linear low density polyethylene. Results Phys 8:461–467
11. Fuqua MA, Huo S, Ulven CA (2012) Natural fiber reinforced composites. Polym Rev 52(3):259–320. https://doi.org/10.1080/15583724.2012.705409

12. Pereira A, Monteiro S, Assis F, Colorado H (2017) Charpy toughness behavior of fique fabric reinforced polysester matrix composites. In: Characterization of minerals, metals, and materials 2017. Springer, Cham, pp 3–9
13. Muñoz-Velez MF, Hidalgo-Salazar MA, Mina-Hernandez JH (2014) Fibras De Fique Una Alternativa Para El Reforzamiento De Plásticos. Influencia De La Modificación Superficial. Biotecnologia En El Agropecuario Y Agroindustrial 12(2), 60–70
14. Tonoli G, Santos S, Savastano H Jr, Delvasto S, Gutiérrez RM, Murphy MD (2011) Effects of natural weathering on microstructure and mineral composition of cementitious roofing tiles reinforced with fique fibre. Cement Concr Compos 33:225–232. https://doi.org/10.1016/j.cemconcomp.2010.10.013
15. Hoyos CG, Vázquez A (2012) Flexural properties loss of unidirectional epoxy/fique composites immersed in water and alkaline medium for construction application. Compos Part B 43: 3120–3130. https://doi.org/10.1016/j.compositesb.2012.04.027
16. Hoyos CG, Alvarez VA, Rojo PG, Vázquez A (2012) Fique fibers: enhancement of the tensile strength of alkali treated fibers during tensile load application. Fibers Polym 13(5):632–640. https://doi.org/10.1007/s12221-012-0632-8
17. Luna P, Mariño A, Lizarazo-Marriaga J, Beltrán O (2017) Dry etching plasma applied to fique fibers: influence on their mechanical properties and surface appearance. Procedia Eng 200:141–147. https://doi.org/10.1016/j.proeng.2017.07.021
18. Hidalgo-Salazar MA, Mina JH, Herrera-Franco PJ (2013) The effect of interfacial adhesion on the creep behaviour of LDPE–Al–Fique composite materials. Compos Part B 55: 345–351. https://doi.org/10.1016/j.compositesb.2013.06.032
19. Pereira AC, Monteiro SN, Assis FS, Colorado HA (2017) Izod impact tests in polyester matrix composites reinforced with fique fabric. In: Proceedings of the 3rd Pan American Materials Congress. Springer, Cham, pp 365–372
20. Monteiro SN, De Assis FS, Ferreira CL, Simonassi NT, Weber RP, Oliveira MS, Pereira AC (2018) Fique fabric: a promising reinforcement for polymer composites. Polymers 10(246):10. https://doi.org/10.3390/polym10030246
21. Teles MC, Altoé GR, Netto PA, Colorado H, Margem FM, Monteiro SN (2015) Fique fiber tensile elastic modulus dependence with diameter using the Weibull statistical analysis. Mater Res 18(Suppl 2):193–199
22. Ovalle-Serrano SA, Blanco-Tirado C, Combariza MY (2018) Exploring the composition of raw and delignified Colombian fique fibers, tow and pulp. Cellulose 25(151). https://doi.org/10.1007/s10570-017-1599-9

Chemical and Morphological Characterization of Guaruman Fiber

Raphael Henrique Morais Reis, Verônica Scarpini Cândido, Larissa Fernandes Nunes and Sergio Neves Monteiro

Abstract Guarumã or guaruman fiber is a viable alternative, compared to the synthetic fiber, as reinforcement in composite. Differentiated properties like mechanical strength, biodegradability, and renewable resource are reasons of interest for the development of materials with this fiber. Guaruman fibers (Ishinosiphon Koern) can be found in the regions of floodplain in large quantities and, therefore, is a potential raw material as natural fiber. In this work, guaruman fibers were analyzed using Fourier-transform infrared spectroscopy (FTIR) and scanning electron microscopy (SEM). The results indicated that this fiber showed structures in rectangular forms.

Keywords Guaruman fiber · Chemical characterization · Morphological characterization

Introduction

The natural lignocellulosic fiber (NLF) has been gaining prominence in scientific research in Brazil and the world. Natural fibers can play an important role as reinforcement in composite to replace synthetic fibers. These fibers have low cost, good biodegradability, low toxicity, and low density [1–7].

In previous works, the NLFs had shown superior properties in several areas [8–18]. Some important researches have shown natural fibers can be incorporated in multilayer amor systems (MAS), cement, as well as in the automotive components [1–3]. Because the interest in sustentable products has increased, NFLs' study has recently showed higher priority in materials research [3]. The chemical and microstructural characterization does provide essential information about the composition and fibrillar structure of fibers. This data can indicate what will be the physical and chemical

R. H. M. Reis (✉) · L. F. Nunes · S. N. Monteiro
Military Institute of Engineering—IME, Rio de Janeiro, Brazil
e-mail: raphael.reis.a@gmail.com

V. S. Cândido
Universidade Federal do Pará—UFPA, Belém, Brazil

S. Ikhmayies et al. (eds.), *Green Materials Engineering*, The Minerals, Metals & Materials Series, https://doi.org/10.1007/978-3-030-10383-5_12

properties of related composite and, consequently, the final mechanical properties and interaction between the polymeric matrix and the fiber [4, 5].

Silva et al. [6], investigated the effect of fiber morphology on the tensile strength. Weibull statistical analysis indicated that each fiber has a characteristic morphology, being two groups of fiber classified in high-performance natural fibers and low-performance natural fibers. In fact, the internal area of the lumens together with the secondary cell-wall thickness affects the mechanical properties.

In another NLF, Luz et al. [7], indicated stronger adhesion of PALF with epoxy resin in comparison with the coir fiber. In this study, it was verified a possible justification for the distinct morphological aspects association with the PALF fiber has rougher surface.

The present work presents a systematic study on the chemical and morphological characterization of guaruman fiber to contribute with the several researches about NLFs. The guaruman fiber is the potential raw material to composites, found from north region of Brazil.

Materials and Methods

The guaruman fibers were supplied by Federal University of Pará, Brazil. Their fibers were cut into smaller samples. Small amount of guaruman fibers were sent to FTIR and SEM analyses.

The FTIR analyses were carried out at Chemical Department of the Military Engineering Institute (IME). To the authors' knowledge, no articles with FTIR analyses to guaruman fiber are available. Table 1 shows the relationship between the band origin and positions in the NLFs [19, 20].

SEM investigations were carried out using a FEI Quanta 250 FEG-SEM at Materials Science Department of IME. Chemical composition analyses were obtained using the attached EDS system.

Results and Discussion

Guaruman fibers showed structures with rectangular forms, which are different from the majority of fibers with circular patterns. This can result in different mechanical properties of the produced materials compared to others fibers, such as the interface interaction between polymeric matrix and natural fiber.

Figure 1 shown the result of the SEM images with magnification 100×. In first view, it is noted that the fiber is formed by continuous fibrils, when looked with more attention it can be observed that some fibrils seem to form scales. It's possible that scales fibrils be the morphological characteristic form theirs.

The continuous fibers have been rough in appearance and that can be an indication of low-interfacial interaction between fiber and matrix. The scales are distributed

Table 1 FTIR adsorption bands of lignin [19, 20]

Position (cm^{-1})	Band origin
3450-3400	O–H stretching
3050-2840	C–H stretching (aliphatic+aromatic)
1740-1710	C=O stretching (unconjugated ketone, ester or carboxylic groups)
1675-1660	C=O stretching in conjugation to aromatic ring
1605-1600	Aromatic ring vibrations
1515-1505	Aromatic ring vibrations
1470-1460	C–H deformations
1430-1425	Aromatic ring vibrations
1370-1365	C–H deformations
1330-1325	Syringyl ring breathing
1275-1270	Guaiacyl ring breathing
1230-1220	C–C, C–O stretch
1172	C–O stretching of conjugated ester groups in grass lignins
1085-1030	C–H, C–O deformations
835	C–H out of plane in p-hydroxyphenyl units

Fig. 1 Scanning electron microscopy (SEM) image with magnification of 100×

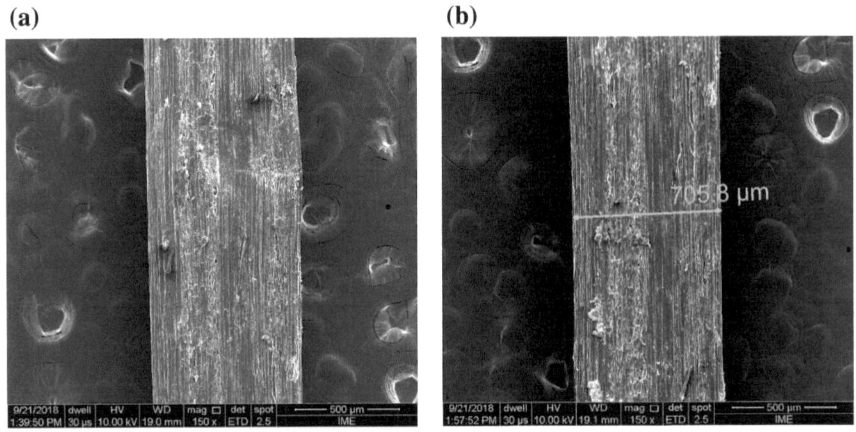

Fig. 2 Scanning electron microscopy (SEM) image with magnification of 150× **a** without measure and **b** with measure

along the fiber and change their interaction. Because of small opening, there is a chance for polymeric matrix to adhere to the fiber.

Figure 2a and b show SEM images with magnification of 150×. Observing the Fig. 2a is important to note that remains more clear the form of scale fibrils. However, it was also found that the other possible fiber's microstructure characteristics are similar to the fiber outside the matrix.

Some other aspects of guaruman fiber were evidenced in these images. Firstly, in some places of the fiber, there are fibrils with fracture potential. This is because the continuous fibers in certain areas were easily broken. Also, it is reasonable to say that some fibril tips are highlighted, which can also be an aspect characteristic of this fiber.

Figure. 2b was measured the size of the guaruman fiber with approximately 705.8 μm of width. Moreover, it was measured with a digital caliper and its thickness of approximate 80 μm. Thus, it was confirmed that this fiber has a rectangular cross section.

Figure 3a and b correspond the SEM microscopy image with magnitude of 500× and 1000×, respectively. These images seek to visualize the morphological characterization on the guaruman fiber.

In these figures, it can be observed that the fibrils stand out like mini cylinders. Part of the fibrils folded into a fibrillar organization. As a result, in the Fig. 3a fibrils can be seen folded in a stick shape. Meanwhile, in the Fig. 3b fibrils can be seen folded in cylinders. Therefore, probably the stick and cylinder shapes are the morphological characteristics of the guaruman fibers.

Figure 4 shows the spectra found for the FTIR from guaruman fiber. In this figure, one can see several peaks that represent the wave number of an absorption band. That is, each wave number sets up a group of the lignocellulosic fiber composition. Table 1 can analyze the relationship of wave number with lignocellulosic fiber composition.

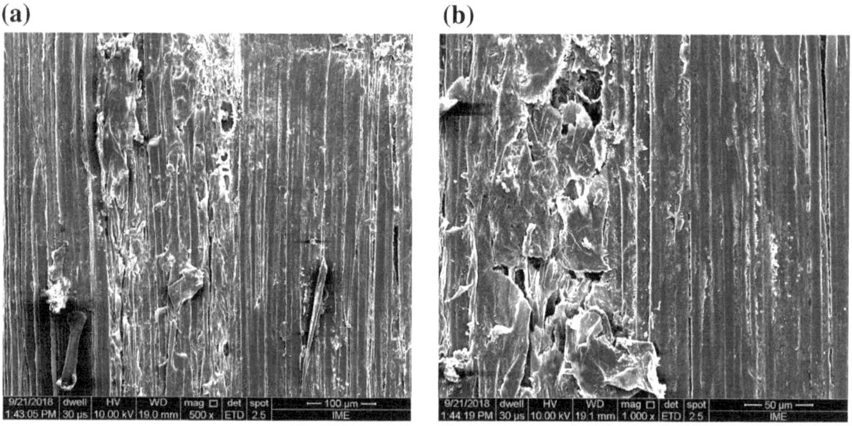

Fig. 3 Scanning electron microscopy (SEM) image with magnification of 500× (**a**) without measure and 1000× (**b**)

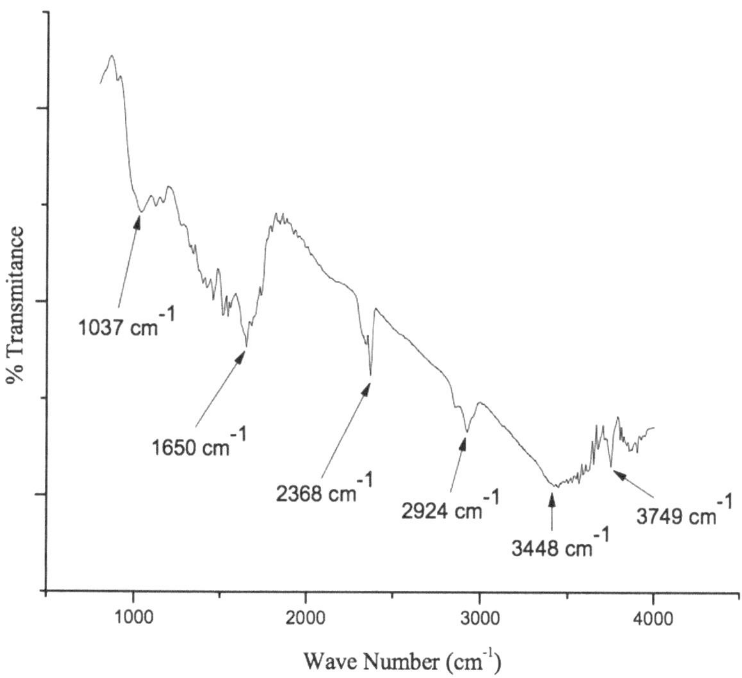

Fig. 4 FTIR spectra of guaruman fiber

FTIR spectra showed the absorption bands of chemical group characteristics to guaruman fibers: cellulose, hemicellulose, and lignin. The absorptions point to the following groups: (I) in position 3448 cm^{-1} could be corresponding the band of O–H stretching; (II) the band 2924 cm^{-1} can be ascribed to C–H stretching (aliphatic+aromatic); (III) in the position around 1650 cm^{-1} the band is due to C=O stretching in conjugation to aromatic ring; (IV) the wave number 1037 cm^{-1} possibility be the C–H, C–O deformations, and (V) the band 3749 and 2368 cm^{-1} do not have corresponding position from Table 1.

From literature, the band of O–H stretching in 3448 cm^{-1} is characteristic of the hydroxyl groups present in the cellulose, hemicellulose, water, and lignin structures, beside extractives and carboxylic acids. The absorption band around 2924 cm^{-1} is characteristic in any macromolecule present in the cellulose and hemicellulose components existing in natural fiber [19–22]. The band 1650 cm^{-1} corresponds to the carbonyl (C=O) stretching vibration of aliphatic carboxylic acids and ketones, mainly due to hemicellulose groups [19–23]. The characteristic band 1037 cm^{-1} is also often found in NLFs, that is, polysaccharide components mainly of cellulose.

Summary and Conclusions

- Guaruman fibers have two possibly specific morphological characteristics. Firstly, fibrils would be folded in the stick shape. Secondly, the fibrils would be folded in cylindrical shape. Also, it can be considered that most of the morphological characteristics of guaruman fibers are of continuous and slick appearance. In addition, guaruman fiber exhibited a rectangular cross section demonstrated with the measure by SEM and digital caliper.
- FTIR spectrum for the guaruman fiber shows similar bands usually found in NLFs. Furthermore, the absorptions point at 2368 and 3749 cm^{-1} haven't been present characteristic bands typical of cellulign materials. Indeed, the wave number 2368 cm^{-1} was related in a previous article about Malva fiber [20]. However, the peak 3749 cm^{-1} was not reported for other NLFs until the present study.

Acknowledgements The authors thank the Brazilian agencies CAPES, FAPERJ, and CNPq for the financial support.

References

1. Nascimento DCO, Lopes FPD, Monteiro SN (2010) Tensle behavios of lignocellulosic fiber reinforced polymer composites: part I jute/epoxy. Rev Mater 15(2):199–205
2. Colorado HA, Quintero-Dávila M, Monteiro SN (2018) Composites of portland cement and fibers of Guadua angustifolia kunth from Colombia. Compos Mater 0(0): 1–10
3. Cruz RBD, Junior L, Pereira E, Monteiro SN, Louro LHL (2015) Giant bamboo fiber reinforced epoxy composite in multilayered ballistic armor. Mater Res 18:70–75

4. Li X, Tabil LG, Panigrahi S (2006) Chemical treatments of natural fiber for use in natural fiber-reinforced composites: a review. J Polym Environ 15:25–33
5. Lansing AI, Rosenthal TB, Alex M, Dempsey EW (1952) The structure and chemical characterization of elastic fibers as reveled by elastase and by electron microscopy. Anat Rec 114(4):555–575
6. Silva FA et al (2013) The effect of fiber morphology on the tensile strength of natural fibers. Mater Res Technol 2(2):149–157
7. Luz FS, Ramos FJHTV, Nascimento LFC, Figueiredo ABHS, Monteiro SN (2018) Critical length and interfacial strength of PALF and coir fiber incorporated in epoxy resin matrix. JMRET, 1–7
8. Crocker J (2008) Natural materials innovative natural composites. Mater Technol 23(3):174–178
9. John MJ, Thomas S (2008) Biofibers and biocomposites. Carbohydr Polym 71(3):343–364
10. Monteiro SN, Lopes FPD, Ferreira AS, Nascimento DCO (2009) Natural fiber polymer matrix composites: cheaper, tougher and environmentally friendly. JOM 61:17–22
11. Monteiro SN, Lopes FPD, Barbosa APB, Bevitori AB, Da Silva ILA, Da Costa LL (2011) Natural lignocellulosic fibers as engineering materials—an overview. Metall Mater Trans A 42A:2963–2974
12. Faruk O, Bledzki AK, Fink HP, Sain M (2012) Biocomposites reinforced with natural fibers. Prog Polym Sci 37(11):1555–1596
13. Thakur VK, Thakur MK, Gupta RK (2014) Review: raw natural fibers based polymer composites. Int J Polym Anal Charact 19(3):256–271
14. Guven O, Monteiro SN, Moura EAB, Drelich W (2016) Re-emerging field of lignocellulosic fiber-polymer composites and ionizing radiation technology in then formulation. Polym Rev 56:702–736
15. Pickering KL, Effendy MGA, Le TM (2016) A review of recent developments in natural fibre composites and their mechanical performance. Compos Part A Appl Sci Manuf 86:98–112
16. Holbery J, Houston D (2006) Natural fiber reinforced polymer composites in automotive applications. JOM 58(11):80–86
17. Zah R, Hischier R, Leal AL, Brown I (2007) Curaua fibers in automobile industry: a sustainable assessment. J Clean Prod 15(11–12):1032–1340
18. Trindade WG, Paiva JMF, Leão AL, Frollini E (2008) Ionized-air-treated curaua fibers as reinforcement for phenolic matrices. Macromol Mater Eng 293(6):521
19. Monteiro SN et al (2014) Characterization of banana fibers functional groups by infrared spectroscopy. Mater Sci 775:250–254
20. Monteiro SN et al (2014) Infra-red spectroscopy analysis of malva fibers. Mater Sci 775:255–260
21. De Paoli MA et al (2009) Characterization of lignocellulosic curaua fibres. Carbohydr Polym 77:47–53
22. Romanzini D et al (2012) Preparation and characterization of ramie-glass fiber reinforced polymer matrix hybrid composites. Mater Res 15(3):415–420
23. Satyanarayana KG et al (2009) Characterization of banana, sugarcane bagasse and sponge gourd fibers of Brazil. Ind Crop Prod 30:407–415

Part IV
Properties and Characterization of Green Materials

Characterization of Arapaima Fish Scales and Related Reinforced Epoxy Matrix Composites by XRD, EDS, and SEM

Wendell B. A. Bezerra, Michelle S. Oliveira, Fabio C. Garcia Filho, Luana C. C. Demosthenes, Luís Carlos da Silva and Sergio N. Monteiro

Abstract Arapaima gigas, also known as pirarucu in Brazil, is one of the biggest freshwater fishes. The structure of the Arapaima scales is a composite consisting of a mineralized outer layer with surface corrugations and a flexible internal collagenous foundation and can be used as an inspiration for developing flexible composites as an intermediate layer in multilayered armor system. In this paper, the structure of the Arapaima scales was reviewed and structural, chemical, and morphological properties of the reinforced composites with 30% vol. of Arapaima scales. The SEM analysis of the morphology of the scales confirmed a sandwich structure, composed of an internal porous matrix surrounded by two external dense layers. Chemical analysis by XRD revealed that the scales' outer surface contains calcium-deficient hydroxyapatite. The EDS results confirm that the percentage of calcium is higher in the outer layer. The experimental results provide an initial view of the potential use of Arapaima scales in epoxy composites.

Keywords Arapaima gigas · Armor · Pirarucu · Morphology

W. B. A. Bezerra (✉) · M. S. Oliveira · F. C. Garcia Filho · L. C. C. Demosthenes · L. C. da Silva · S. N. Monteiro
Military Institute of Engineering—IME, Rio de Janeiro, Brazil
e-mail: wendellbez@gmail.com

S. Ikhmayies et al. (eds.), *Green Materials Engineering*, The Minerals, Metals & Materials Series, https://doi.org/10.1007/978-3-030-10383-5_13

Introduction

Some natural shielding materials, such as alligator osteoderms, turtle, and armadillo shells and fish scales, have developed and improved their purposes over millions of years through a process of convergent evolution [1]. The study of biological materials has been increasingly attractive and reveals a range of new structures, internal design, and mechanical responses that even and often outnumber synthetic materials of similar functions [2].

Arapaima gigas, also known as pirarucu in Brazil, is considered one of the biggest freshwater fish, being able to reach a maximum weight of 200 kg and length between 2 and 3 m [3]. These fish are covered by scales that may reach 10 cm in length which acts as an auxiliary mean of self-defense, due to their natural habitat being the rivers of the Amazon basin, which become infested by piranhas with the advance of the dry season [2, 4].

Arapaima scales are an example of material that has evolved for better protection against predators [2]. The scales present laminate structure formed by collagen fibers mineralized in a plywood-like pattern and reinforced by hydroxyapatite nanocrystals [5]. Collagen fibers are ordered in a cross-lamellar arrangement that results in a laminated composite [2]. Figure 1 shows a schematic representation of the structure of the Arapaima scales.

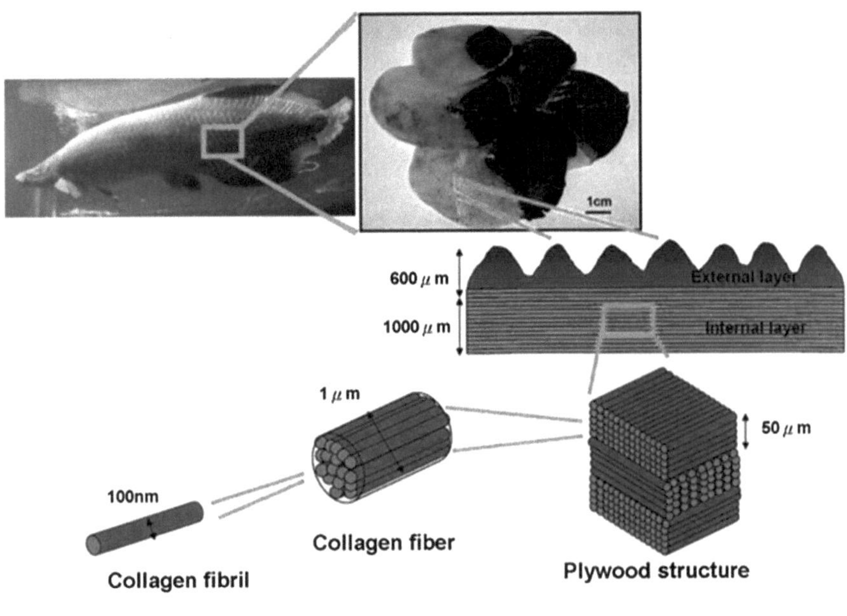

Fig. 1 Representation of the hierarchical structure of the Arapaima gigas scales [4]

Several studies have been devoted to investigate the structure and properties of Arapaima scales in recent years. Torres et al. [5] studied the structure and composition of the scales through XRD and FTIR analysis, confirming the presence of hydroxyapatite and collagen. Sherman et al. [6] and Murcia et al. [7] carried out comparative studies of the structures found in the Arapaima scales with those observed in other species of fish. Torres et al. [8] analyzed the influence of the presence of varying water contents on the thermal transition of the scales through differential calorimetry analyzes. Arola et al. [9] studied the structure, composition, and mechanical behavior of the boundary layer of scales for three different fish, namely pirarucu (Arapaima gigas), tarpon (Megalops atlanticus), and carp (Cyprinus carpio).

The development of composite materials has become increasingly common and simple through the use of fibers as reinforcement in polymer matrices. Factors such as the high price of synthetic fibers and the need to use renewable and non-polluting resources are great motivators for the use of natural fibers instead of traditional ones [10]. In this context, the Arapaima scales present themselves as a potential alternative for the use as reinforcement in polymer matrices.

Thus, the objective of this work is to characterize the structure, morphology, and composition of Arapaima scales and 30% vol. epoxy/Arapaima scales composites through scanning electron microscopy (SEM), energy-dispersive X-ray spectroscopy (EDS), and X-ray diffraction (XRD).

Experimental Procedure

Arapaima (Arapaima gigas) scales from the state of Para, Brazil, were provided by one of the co-authors [LCCD]. Figure 2 shows as-received Arapaima scales. These scales were cleaned in running water and dried under pressure to become flat.

The surface morphology of the samples was studied by scanning electron microscopy (SEM) with energy-dispersive X-ray spectroscopy (EDS) on both FEI Quanta FEG 250 and Hitachi TM3000, using an acceleration voltage of 20 kV. The EDS allowed qualitative identification of the chemical elements present in the samples.

In order to make possible the visualization of the surfaces of the scales, it was necessary to use a conductive coating of platinum, for which a high-vacuum film deposition equipment of the brand LEICA (model EM ACE600) was used.

The X-ray diffraction (XRD) analysis was performed at room temperature in an X'Pert Pro Panalytical diffractometer, operating at a 40 kV voltage with a current of 40 mA, the scan angle ranged from 10° to 70° with step sizes of 0.05° and CuKα radiation was used.

Fig. 2 Arapaima scales as-received

Results

Figure 3 shows the SEM images obtained for Arapaima gigas scales at different regions and magnifications. By observation of the images, it is possible to see a variation in the topography of the Arapaima scales surface, with the presence of a rough texture in some regions and smooth in others. Voids can be noticed over the surface at Fig. 3a and d which reveals a laminated structure, as observed in other studies [2, 5]. It is possible to notice the presence of a fibrous-like zone near the voids that may be associated with the veins of the scale in Fig. 3a, as reported in the previous studies [2].

Figure 4 presents the results obtained by the EDS analysis for different regions observed of Arapaimas scales. In Fig. 4a, it is possible to observe the absence of Ca and Si in the composition of the scales, compared to the other analyzed areas of

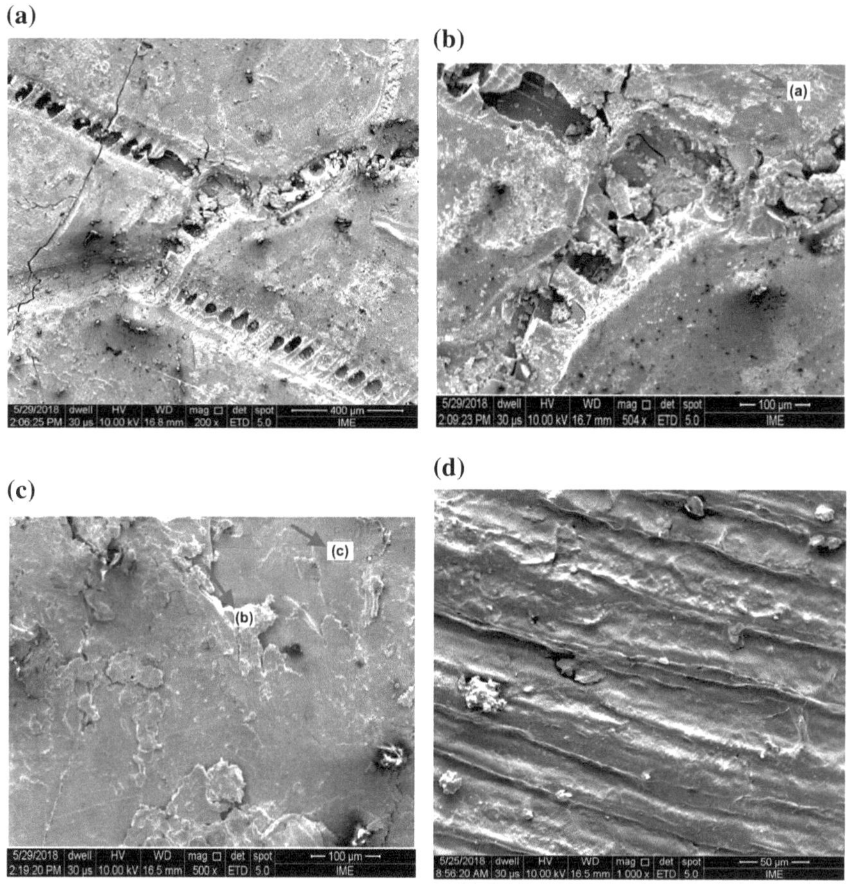

Fig. 3 Photomicrographs of the surface of the Arapaima scales in different regions and magnifications: **a** 400x, **b** 504x, **c** 500x, and **d** 1000x

the scale, b and c. This phenomenon can be explained by the occurrence of a higher mineralization of these regions, which contributes to the formation of voids with a higher concentration of Ca, which is observed in Fig. 4b and c [4, 11]. While regions with less mineralization, such as the one analyzed in Fig. 4a, they present filled pores with a higher concentration of C and O and smaller amounts of Ca [4, 11].

Figure 5 shows the diffractogram obtained for the Arapaima scales, where it is possible to observe the presence of peaks at $2\theta = 26$; 31.9; 39.6; 49.8, and 53.2°, similar to those obtained in other studies [4, 5]. The presence of broad peaks corresponding to the apatite structure reveals that the scales have relatively low levels of crystallinity [5].

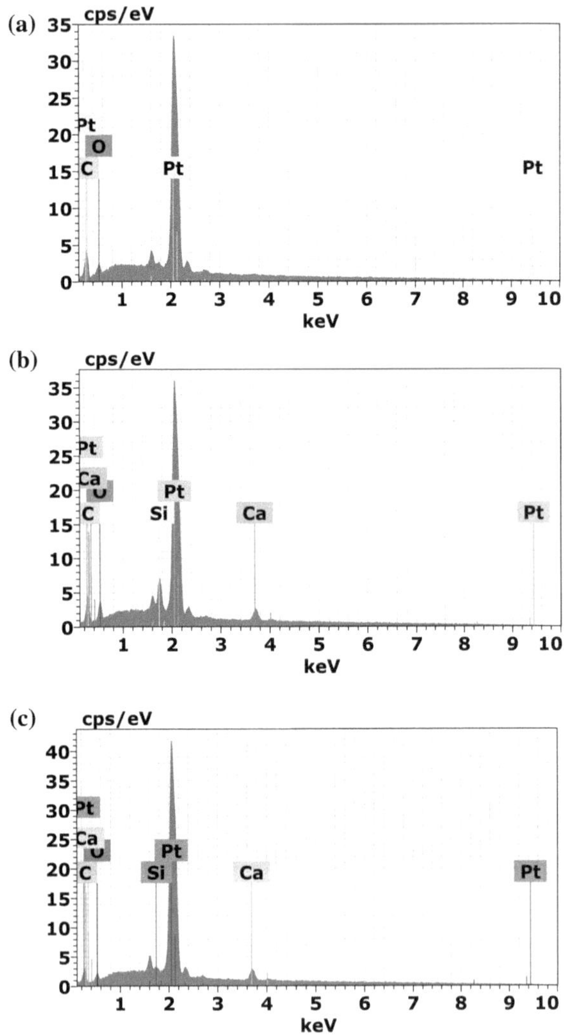

Fig. 4 EDS results for the surface of the Arapaima scales from the regions shown in Fig. 2a, b, and c

Figure 6 shows the diffractogram obtained for the reinforced composite with 30% by volume of Arapaima scales. It is possible to observe the presence of the peak associated with the scales at $2\theta = 31.9°$, while the amorphous halo at lower angles is associated with the epoxy resin used. These results might indicate that there was a good adhesion between the scales and the epoxy matrix.

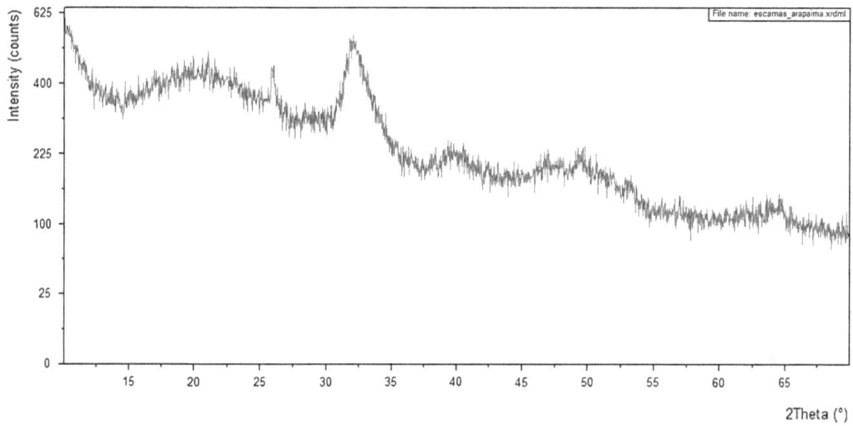

Fig. 5 XRD spectra of the Arapaima scales

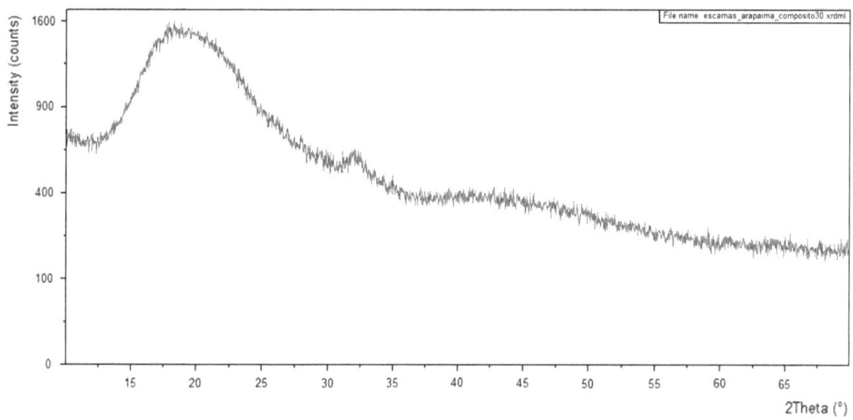

Fig. 6 XRD spectra of composites reinforced with 30% vol. of Arapaima scales

Summary and Conclusions

- The wide peaks of the DRX pattern indicate that the scale has low crystallinity and it is possible to observe the amorphous halo in the range of $2\theta = 10–30°$ corresponding to the epoxy matrix in the composite.
- The EDS results showed that the calcium content is responsible for the scale mineralization, which leads to a higher hardness. The peak related to platinum (Pt) refers to the conductive coating utilized.
- SEM observation of Arapaima scales reveals layers, so they can be thought of as-laminated composite structures formed by mineralized collagen fibers in a pattern of plywood layers.

References

1. Yang W et al (2014) Protective role of arapaima gigas fish scales: structure and mechanical behavior. Acta Biomater 10:3599–3614
2. Meyers MA et al (2012) Battle in the amazon: arapaima versus piranha. Adv Eng Mater 14(5):279–288
3. Imbiriba EP (2001) Potencial de criação de pirarucu, Arapaima gigas, em cativeiro. Acta Amaz 31(2):299–316
4. Meyers MA et al (2011) Mechanical properties and the laminate structure of arapaima gigas scales. J Mech Behav Biomed Mater 4(7):1145–1156
5. Torres FG et al (2008) Characterization of the nanocomposite laminate structure occurring in fish scales from arapaima gigas. Mater Sci Eng C 28:1276–1283
6. Meyers MA et al (2017) A comparative study of piscine defense: the scales of arapaima gigas, latimeria chalumnae and atractosteus spatula. J Mech Behav Biomed Mater 73:1–16
7. Murcia S et al (2017) The natural armors of fish: a comparison of the lamination pattern and structure of scales. J Mech Behav Biomed Mater 73:17–27
8. Torres FG, Troncoso OP, Amaya E (2012) The effect of water on the thermal transitions of fish scales from arapaima gigas. Mater Sci Eng C 32:2212–2214
9. Arola D et al (2018) The limiting layer of fish scales: structure and properties. Acta Biomater 2018(67):319–330
10. Barbosa AP (2011) Structural characteristics and properties of polymeric composite reinforce with buriti fibres. PhD thesis, UENF
11. Ebenstein D, Calderon C, Troncoso OP, Torres FG (2015) Characterization of dermal plates from armored catfish pterygoplichthys pardalis reveals sandwich-like nanocomposite structure. J Mech Behav Biomed Mater 45:175–182

Piassava Fibers: Morphologic and Spectroscopic Aspects

Fabio da Costa Garcia Filho, Luana Cristyne da Cruz Demosthenes, Michelle Souza Oliveira, Artur Camposo Pereira, Fernanda Santos da Luz and Sergio Neves Monteiro

Abstract The use of natural lignocellulosic fibers (NLFs) as reinforcement in polymer matrix composites is being considered by many different industries, for instance, the automotive industry where components such as interior panel and cushion seat are being manufactured using NLFs. The piassava fiber is a less characterized NLF when compared to sisal or jute, although the properties displayed by the piassava fiber make them suitable as reinforcement to composite materials. Therefore, this work aimed to present a morphological analysis of the piassava fiber by scanning electron microscopy (SEM) as well as studied the spectroscopy aspects of this fiber by Fourier transform infrared (FTIR). The SEM analysis revealed differences in their shape and size as well as a rough aspect associated with a great number of protrusions. While the FTIR analysis showed a hydrophilic character of the fibers due to the presence of axil vibration of hydroxyl groups band between 3500 and 3300 cm^{-1}.

Keywords Natural fiber · Morphology · FTIR

Introduction

In recent years, it has been an increasing effort to reduce the amount CO_2 emission which can be associated with greenhouse effect and global warming [1]. Therefore, an interest in the substitution of green materials for those of synthetic origin has been growing. Natural fibers can be considered as green materials once they exhibit characteristics such as renewability and recyclability but also biodegradability [2, 3]. Moreover, abundance as well as low cost of production are considered as the major advantages that motivate not only researchers but also industries to invest in the use of these materials [2, 4, 5]. It has been reported that these fibers present mechanical properties compared to those exhibit by the glass fibers [6]. And when

F. da Costa Garcia Filho (✉) · L. C. da Cruz Demosthenes · M. S. Oliveira · A. C. Pereira · F. S. da Luz · S. N. Monteiro
Department of Materials Science, Military Institute of Engineering - IME, Praça General Tibúrcio, 80, Praia Vermelha, Urca, Rio de Janeiro, RJ 22290-270, Brazil
e-mail: fabiogarciafilho@gmail.com

S. Ikhmayies et al. (eds.), *Green Materials Engineering*, The Minerals, Metals & Materials Series, https://doi.org/10.1007/978-3-030-10383-5_14

Fig. 1 Piassava palm tree (**a**) and bundle of piassava fibers (**b**)

used as reinforcement in polymeric matrix composites can be even applied in ballistic armor systems [7–13].

Many different fibers have been extensively discussed in the literature such as curaua [14] and jute [15] as well as the fibers obtained from the coconut shell [16] and sisal [17]. Yet, there are many other less studied fibers with great potential of being used as engineering materials. Piassava fibers ("*Attalea funifera*") are in that class of promising materials to reinforce polymeric matrices to produce stiffer and tougher composites. The piassava fibers are extracted from palm trees that are grown in the Atlantic rain forest in the southeast region of Brazil. Figure 1a presents the palm tree from where the fibers are extracted while in Fig. 1b the collected fibers are presented as bundles.

Some properties of these fibers have been reported such as density, chemical composition, and mechanical behavior. Density was measured varying in the range of 1.10–1.45 g/cm^3 [18]. High amount of cellulose (>30%) and lignin (>45%) were found in the chemical composition, which may be associated with the water resistance of this fiber [19]. Average tensile strength about 120 MPa, yield strength of approximately 75 MPa and Young's modulus of 5.5–6.3 GPa were calculated for fibers with diameter varying from 0.25 to 0.70 mm [20].

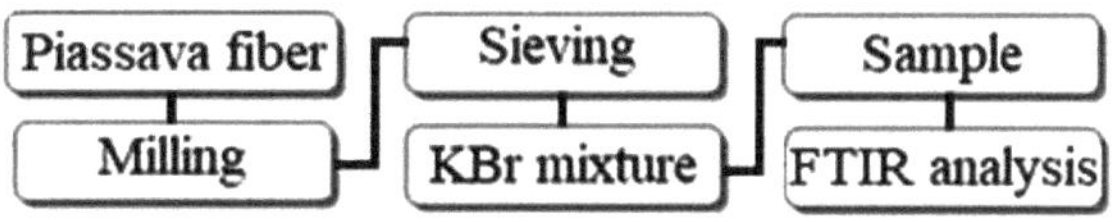

Fig. 2 Procedure for FTIR sample preparation

Nascimento et al. [21] investigated the influence on the use of piassava fibers as reinforcement in epoxy matrix composites. Tensile and flexural strength were analyzed for conditions with 10, 20, 30, and 40% of fiber content as well as a neat condition used as reference. It was observed that the 10% fiber content condition tends to exhibit a lower value of resistance than those of the reference condition. Furthermore, as the amount of reinforcement increased, it was reported a trend in the increasing of the strength and has been even comparable to the rule of mixtures.

More recently, Garcia Filho and Monteiro [22] published a paper discussing the application of epoxy matrix composites reinforced with piassava fibers for ballistic armor systems. It was verified similar mechanisms of kinetic energy absorption as observed for conventional materials such as aramid synthetic fabric as well as an inverse relationship with the amount of fiber used as reinforcement and the number of cracks/fractures after the ballistic impact.

Published literature reveals that, natural fibers, in particular the piassava fiber, have been studied as reinforcement for polymeric matrix composites. On the other hand, many properties of the fiber itself still require further investigation. This research aimed to study the morphological characteristics of the piassava fibers as well as the spectroscopic answer using an infrared radiation (FTIR).

Materials and Methods

Piassava fibers were gently supplied by a broom industry Varrouras Rossi, Brazil, used in this work. The morphological characterization of the fibers was done by scanning electron microscopy (SEM), in a model Quanta FEG 250 FEI equipment. The analysis was performed on gold-sputtered samples using secondary electrons and with a beam voltage of 10–15 kV. The composition of topographic features found at the surface of the fibers was determined by energy dispersive spectroscopy (EDS).

Finally, the characterization by Fourier transform infrared (FTIR) analysis was conducted in a model IR Prestige 21-FTIR-Shimadzu. The piassava fiber sample was prepared in accordance with procedure presented in Fig. 2.

Results and Discussion

Figure 3 shows the typical microscopic appearance of the piassava fiber. In Fig. 3a, the longitudinal surface of the fiber is displayed. One may notice the characteristic

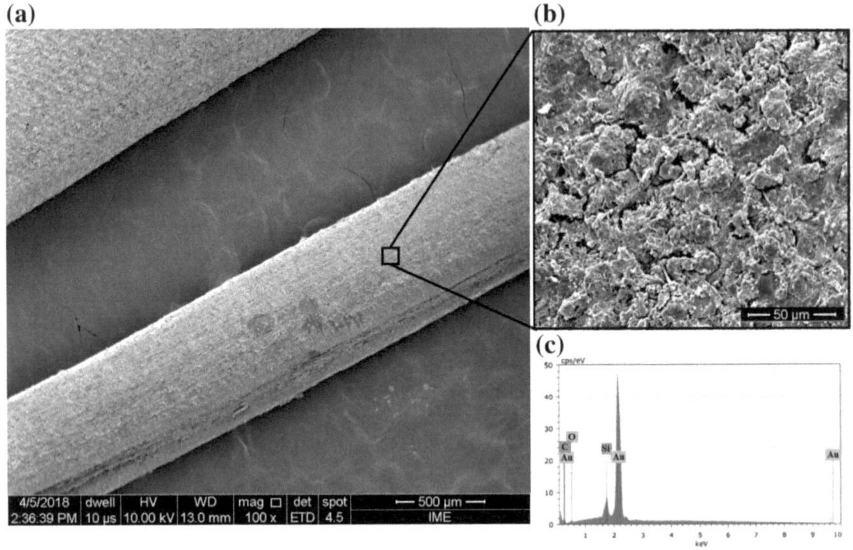

Fig. 3 Microscopic aspect of the piassava fiber (**a**), higher magnification of the piassava surface (**b**) and EDS spectrum of the piassava fiber (**c**)

rough aspect associated with a great number of protrusions. A specific region marked in Fig. 3a is shown in details in Fig. 3b. In this figure, it is possible to verify with greater details the surface protrusions which are Si-rich compounds, as shown in the EDS spectrum exhibit in Fig. 3c.

Figure 4 shows other piassava surface. In Fig. 4a, one should notice the existence of a crack that started at the fiber transverse surface and propagated until the longitudinal surface. This region is marked with a square that is presented with greater detail in Fig. 4b. Figure 4b reveals the rupture of some microfibrils, the fact of only a few of these microfibrils are rupture, suggests that the tensile rupture of the piassava fiber is not an uniform event. This is different from what is reported for sisal and other similar fibers [23].

The investigation of piassava fiber by the Fourier transform infrared spectroscopy (FTIR) technique is essential to reveal the specific molecular component absorption bands in order to understand the interaction that may occur between the piassava fiber and the polymer matrix, when applied as reinforcement in composites. Figure 5 shows the obtained spectrum of FTIR for the piassava fiber.

From this spectrum, an absorption band appears around 3335 cm^{-1}, this can be associated with the axial vibration of hydroxyl groups (O–H) of the cellulose. Khan et al. [24] argue that this vibration of O–H groups can be associated with absorbed water, as well as alcohols found in cellulose, hemicellulose, and lignin. The second absorption band was found at 2912.83 cm^{-1} refers to the vibrational modes of CH_2 and CH_3 molecules, which is a characteristic organic molecular structure of any natural fiber. Da Silva et al. [25] showed that the band around 3400 cm^{-1} may be

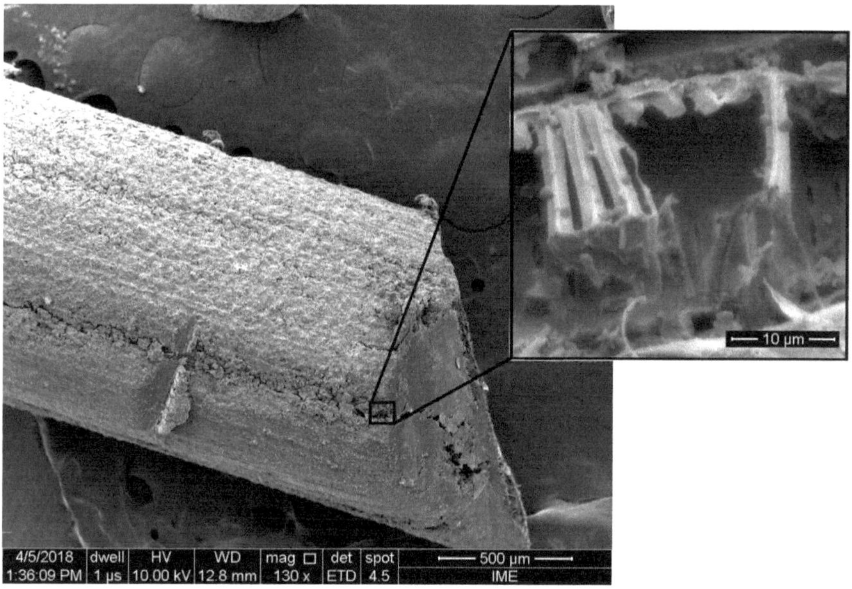

Fig. 4 Piassava longitudinal surface (**a**) and rupture of the micro fibrils in a cracked region (**b**)

Fig. 5 Piassava longitudinal surface (**a**) and rupture of the micro fibrils in a cracked region (**b**)

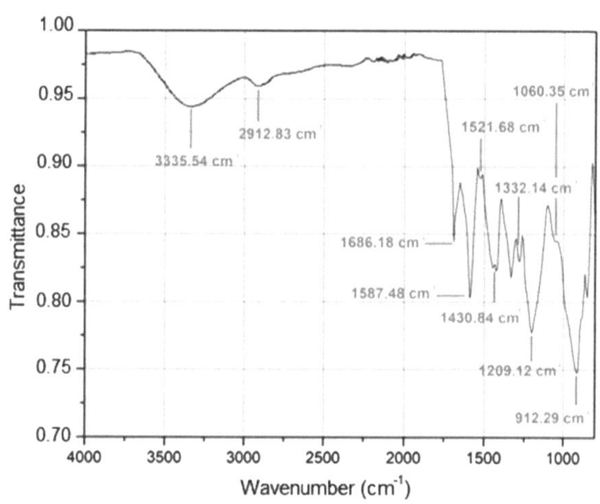

associated with the ease of water desorption of the natural fiber. On the other hand, the band about 2900 cm^{-1} can represent the ease of reaction of the natural fiber components (cellulose, hemicellulose, and lignin), in particular, to the cellulose in interacting with polymeric materials during the manufacture of composites. The band exhibited after 2000 cm^{-1} is associated with the deformation of the bond in the organic molecules. The 1686 cm^{-1} band is attached to the C=O elongation, while the 1587.48 cm^{-1} band is related to the C–O and C–C bonds.

Conclusions

- Morphological as well as spectroscopic aspects of the piassava natural fiber were presented and discussed. The surface morphology analysis was performed using scanning electron microscopy, and the specific molecular absorption band was investigated using the FTIR analysis.
- The longitudinal surface of the piassava fiber reveals that the characteristic roughness of these fibers may be associated with the existence of silicon-rich compound, called protrusions. Moreover, the analysis of a fractured region showed that the rupture of microfibrils tends to occur in a non-uniform pattern. This behavior is different from those reposted for other natural fibers, such as the sisal one.
- The FTIR spectrum revealed bands associated with rotation as well as deformation in the molecular structure of the piassava fiber. Bands with higher wavenumber tend to be related with the vibration of the molecule; therefore the bands about 3300 and 2900 cm^{-1} were presented as vibrations of O–H, CH_2, and CH_3 molecular groups. While the bands from 2000 cm^{-1} were related to the deformation of the bonds C=O, C–O, and C–C.

Acknowledgements The authors thank the support of this investigation by the Brazilian agencies: CNPq, FAPERJ, and CAPES.

References

1. Mohanty AK, Misra M, Drzal DL (2002) Sustainable biocomposites from renewable resources: opportunities and challenges in the green material world. J Polym Environ 10:19–26
2. Monteiro SN, Lopes FPD, Barbosa AP, Bevitori AB, Silva ILA, Costa LL (2011) Natural lignocellulosic fibers as engineering materials–an overview. Metall Mater Trans A 42:2963–2974
3. Satyanarayana KG, Guimarães JL, Wypych F (2007) Studies on lignocellulosic fibers of Brazil. Part I: source, production, morphology, properties and applications. Compos Part A 38:1694–1709
4. Monteiro SN et al (2009) Natural-fiber polymer-matrix composites: cheaper, tougher, and environmentally friendly—An overview. JOM 61(1):17–22
5. Kalia S, Kaith BS, Kaurs I (2011) Cellulose fibers: bio and nano-polymer composites. Springer, New York, USA
6. Wambua P, Ivens I, Verpoest I (2003) Natural fibers: can they replace glass and fibre reinforced plastics? Compos Sci Technol 63:1259–1264
7. Holbery J, Houston D (2006) Natural fiber reinforced polymer composite in automobile application. JOM 60:80–88
8. Pickering K (2008) Properties and performance of natural-fibre composites. CRC Press, Boca Raton, USA
9. Satyanarayana KG, Arigaza GC, Wypych F (2009) Biodegradable composites based on lignocellulosic fibers–an overview. Progr Polym Sci 34:982–1021
10. Güven O, Monteiro SN, Moura EAB, Drelich JW (2016) Re-emerging field of lignocellulosic fiber-polymer composites and ionizing radiation technology in their formulation. Polym Rev 56:702–736
11. Faruk O, Bledzki AK, Fink HP, Sain M (2014) Progress report on natural fiber reinforced composites. Macromol Mater Eng 299:9–26

12. Nabi Sahed D, Jog JP (1999) Natural fiber polymer composites a review. Adv Polym Technol 18:351–363
13. Benzait Z, Trabzon L (2018) A review of recent research on materials used in polymer-matrix composites for body armor application. J Compos Mater 52(23):3241–3263
14. Zah R, Hischier R, Leão AL, Brown I (2007) Curaua fibers in automobile industry – a sustainability assessment. J Clean Prod 15:1032–1040
15. Alves C, Ferrão PMC, Silva AJ, Reis LG, Freitas M, Rodrigues LB (2010) Ecodesign of automotive components making use of natural jute composites. J Clean Prod 18:313–327
16. Luz FS, Monteiro SN, Tommasini FJ (2018) Evaluation of dynamic mechanical properties of PALF and coir fiber reinforcing epoxy composites. Mater Res 21:1–7
17. Senthilkumar K, Saba N, Rajini N, Chandrasekar M, Jawaid M, Siengchin S, Alotman OY (2018) Mechanical properties evaluation of sisal fibre reinforced polymer composites: a review. Constr Build Mater 174(20):713–729
18. Aquino RCMP, D'Almeida JRM, Monteiro SN (2000) Analysis and characterization of piassava fibers. Acta Microsc 9(A): 3–4
19. Schuchardt U, Bianchi ML, Gonçalves AR, Curvelo AAS, Biscolla FC, Peres LO (1995) Piassava fibers (Attalea funifera) I–chemical analysis, extraction and reactivity of its lignin. Cellul Chem Technol 29:705–712
20. Monteiro SN (2008) Properties and structure of Attalea funifera (piassava) fibers for composite reinforcement–a critical discussion. Nat Fibers 6(2):191–203
21. Nascimento DCO, Ferreira AS, Monteiro SN, Aquino RCMP, Satyanarayana GK (2012) Studies on the characterization of piassava fibers and their epoxy composites. Compos Part A 43:353–362
22. Garcia Filho FC, Monteiro SN (2018) Piassava fiber as na epoxy matrix composite reinforcement for ballistic armor applications. JOM. https://doi.org/10.1007/s11837-018-3148-x
23. Mukherjee PS, Satyanarayana KG (1986) An empirical evaluation of structure/property relationships in natural fibers and their fracture behavior. J Mater Sci 21:4162–4168
24. Khan MA, Idriss Ali KM, Basm SC (1993) IR studies of wood plastic composites. J Appl Polym Sci 49:1547–1551
25. Da Silva ILA, Bevitori AB, Rohen LA, Margem FM, Braga FO, Monteiro SN (2016) Characterization of fourier transform infrared (FTIR) analysis for natural jute fiber. Mater Sci Forum 869:283–287

Structural Characterization of Fique Fabric Reinforcing Epoxy Matrix Composites by XRD and SEM Analysis

Michelle Souza Oliveira, Artur Camposo Pereira, Fabio da Costa Garcia Filho, Luana Cristyne da Cruz Demosthenes, Fabio de Oliveira Braga, Fernanda Santos da Luz and Sergio Neves Monteiro

Abstract During the last decades, an increased interest has been reported in the use of polymeric matrix composites reinforced with natural fibers for engineering applications. The advantages on the use of these materials include the replacement of synthetic fibers for natural ones, taking advantage of biodegradable resources and contribute for environment issues. This work aims to evaluate four different fractions of natural fiber reinforcing epoxy matrix composites. The natural fiber used in this investigation was the fique fiber, which due to its properties can be considered a promising candidate for composite reinforcement application. Initially, the fique fabrics were cut into 150 mm × 120 mm dimensions and dipped into a metallic mold. The fractions considered were 15, 30, 40, and 50% of a plain fabric, corresponding to 1, 2, 3, and 4 layers, respectively. The composites were characterized by X-ray diffraction (XRD) and scanning electron microscopy (SEM). The X-ray diffractograms revealed characteristics peaks associated with the fique fiber. It was also showed a relationship between the degree of crystallinity and the amount of reinforcement. Finally, the SEM analysis presented the morphological aspect of the composites as well as the differences associated with the increase of reinforcement.

Keywords Fique fiber · Epoxy composite · Structural characterization

Introduction

Due to the growing concern related to environmental issues and sustainability as well as renewability, new research has been directed at the use of bio-friendly materials.

M. S. Oliveira (✉) · A. C. Pereira · F. da Costa Garcia Filho · L. C. da Cruz Demosthenes · F. S. da Luz · S. N. Monteiro
Military Institute of Engineering – IME, Rio de Janeiro, Brazil
e-mail: oliveirasmichelle@gmail.com

F. de Oliveira Braga
National Service of Industrial Apprenticeship - SENAI, Rio de Janeiro, Brazil

F. de Oliveira Braga
Fluminense Federal University - UFF, Rio de Janeiro, Brazil

S. Ikhmayies et al. (eds.), *Green Materials Engineering*, The Minerals, Metals & Materials Series, https://doi.org/10.1007/978-3-030-10383-5_15

The most discussed sustainability parameters are related to the control of electric energy consumption, greenhouse emissions, and the substitution of traditional materials for alternative materials. This environmental awareness encourages the use of natural fibers as reinforcement in polymer matrices. The use of these fibers has a major impact on the reduction, in the dependence on materials from non-renewable sources, as well as a commitment to environmental and economic issues [1–6]. Moreover, the use of materials from renewable sources has been extensively studied, due to the advantages they offer when compared to synthetic materials, such as high mechanical properties per unit weight, low density, and small manufacturing cost per unit volume [7–9]. An important factor that favors the use of natural fibers as a renewable input is the growing perspective of energy saving by reducing the weight of the components, the recovery of raw materials, and the reuse of materials [10].

In this context, Fique (*Agave Furcraea*) is a Colombian native natural fiber that is a hard fiber extracted by mechanical means from the leaves of a plant. This fiber is a promising candidate for incorporation as polymer matrix composites reinforcement. However, obtaining polymeric composites reinforced with natural fibers requires specific processing conditions, since the natural fibers exhibit a hydrophilic structure, incompatible with hydrophobic thermosetting matrices. In order to the reinforcement be considered efficient, both the matrix and the reinforcement must act together in a given application. Therefore, the interfacial contact has a major importance and must be properly controlled during the processing of the material. Many works indicate that the surface modification of the natural fibers tend to increase the surface energy and improve the compatibility between fiber and matrix [11–17].

The objective of this work was to fabricate and structurally characterize epoxy matrix composites reinforced with fique fibers.

Materials and Methods

Composite plates with dimensions of $120 \times 150 \times 10\ mm^3$ were fabricated with fique fiber fabric and epoxy resin type diglycidyl ether of bisphenol A (DGEBA) hardener with triethylenetetramine (TETA) produced by Dow Chemical and supplied by Epoxyfiber company, Brazil.

The DGEBA/TETA epoxy with stoichiometric ratio of 13 parts of hardener for 100 parts of resin was poured onto the fique fiber fabrics in order to fabricate laminate composites. The reinforcement fractions considered were 15, 30, 40, and 50% of fique fabric, these amounts correspond to 1, 2, 3, and 4 layers, respectively. The final plate was produced by the application of unidirectional load of 3 MPa for 24 h.

The X-ray diffractograms of the plates composites and the fique fiber were obtained from a Shimadzu XRD-6000 X-ray diffractometer, with CuKα radiation source, 40 kV voltage, 40 mA current, and 0.05 (2θ/5 s) sweep for 2θ values between 10 and 40°. The results obtained in the diffractometer allowed to calculate the degree

Table 1 Index of crystallinity of fique fibers

Material	I_{am}	$I_{(002)}$	I_c
Fiber "*in natura*"	1330	2072	35.8

of crystallinity of the fibers through Eq. (1) developed by the empirical method of Segal [10].

$$I_C = \frac{I_{002} - I_{am}}{I_{002}} \quad (1)$$

where I_c is the degree of crystallinity in percentage, I_{002} is the height of the 200 peak representing both crystalline and amorphous materials, or also the peak of diffraction intensity representing the crystalline material near $2\theta = 22°$ and I_{am} the minimum intensity between the 200 and 110 peaks representing only amorphous material near $2\theta = 16°$.

The morphological analysis of the samples was carried out in a scanning electron microscope with tungsten filament operating at 15 kV, using secondary electrons.

Results and Discussion

Based on the XRD patterns of fique fiber and the composites, it was possible to observe the occurrence of two intense peaks, close to the values of $2\theta = 16°$ and 22°. The peak at $2\theta = 16°$ corresponds to the crystallographic plane (101) and the peaks at $2\theta = 22°$. Peak 34° corresponds to the planes (002) and (023) or (004), respectively. According to Hoyos et al. [14], the crystallographic plane (002) corresponds to the native cellulose denominated cellulose I. The XRD (Fig. 1) was performed in order to identify if the incorporation in the composite in different fractions produced any change in the degree of crystallinity. The XRD patterns show changes in signal as a function of fractions.

For fibers with high cellulose percentage, such as cotton, two peaks close to 16° are usually observed. However, in the case of fique fibers, only one peak was found in this region, which is attributed to the presence of amorphous material, such as lignin and hemicellulose, responsible for hiding one of the peaks related to cellulose [5].

Analysis of the X-ray diffraction patterns indicated that the characteristics of the fiber signals with different treatments were similar. However, the peaks for the composites with the greater amount of fiber (Fig. 1b, c, d, and e) were more intense and defined compared to those observed for the "in natura" fibers (Fig. 1a).

The crystallinity index obtained from the fibers is shown in Table 1.

By analyzing the crystallinity index obtained from the fique fiber, one may notice that the value is similar to those reported by other authors [12–15]. Moreover, the degree of crystallinity for the fique is higher than those obtained for palm fiber, about

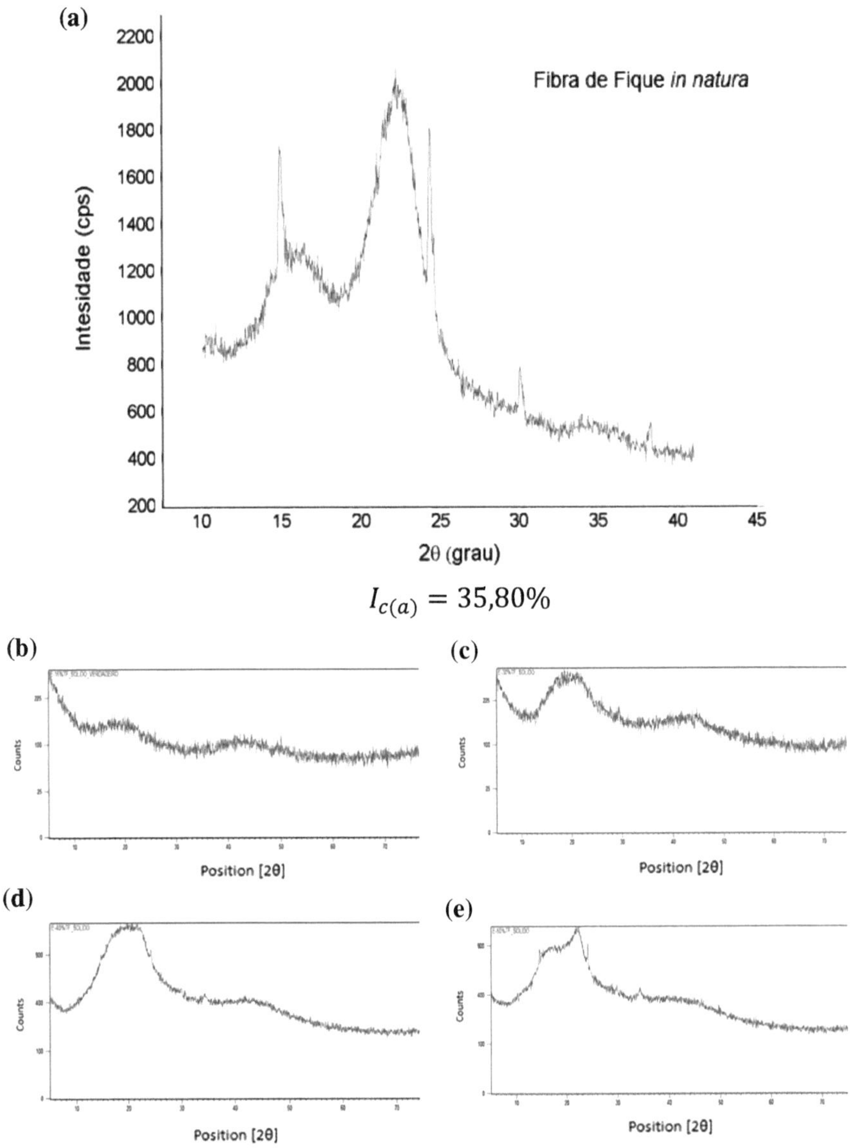

$I_{c(a)} = 35{,}80\%$

Fig. 1 X-ray diffraction of **a** fique fibers and composite with **b** 15%, **c** 30%, **d** 40%, **e** 50% reinforced with fique fibers

Fig. 2 Scanning electronic micrograph: SEM of the fique fibers in natura—helical microfibrils

28%, but lower when compared to other fibers, such as green coconut fibers, about 43%, mallow fiber, about 82.6%, and sugarcane bagasse fibers, which varies in the range of 37.6–58% [2–5].

The aim of the scanning electron microscopy technique was to determine the characteristics of the fibers in terms of morphology and surface appearance. The micrographs of the *in natura* fique fibers show that microfibrils are helically wound along the fiber axis to form ultimate hollow cells, with a spiral angle (angle between fibrils and fiber axis). The fique fiber (Fig. 2) has a porous and homogeneous surface. As sisal fibers and other hard fibers, fique fibers have a hierarchical structure where each fiber is composed of numerous fibrils of 6–30 μm in diameter. These fibrils, held together by hemicellulose and lignin, are easily distinguishable in the micrographs as channels on the fiber's surface [17].

The observation of the interface, as presented in Fig. 3, shows a fique fiber being separated from the matrix. The fiber had been pulled out of the matrix and shows a mild surface with some continuous and slight ridges (Fig. 3a). The matrix presents considerable tearing, it reveals massive failure of both fibers and matrix. The failure mode observed on the fibers shows fiber splitting and tearing, and it is attributed to a satisfactory interaction with the matrix, but still such interaction is by friction and mechanical interlocking.

Furthermore, a composite with the addition of fique fabric in the epoxy matrix with up to 50% less polymer shows viable structural properties for some applications such as ballistic. It's also important to mention that this material also favor its use where cost, environment, and social benefits are considered.

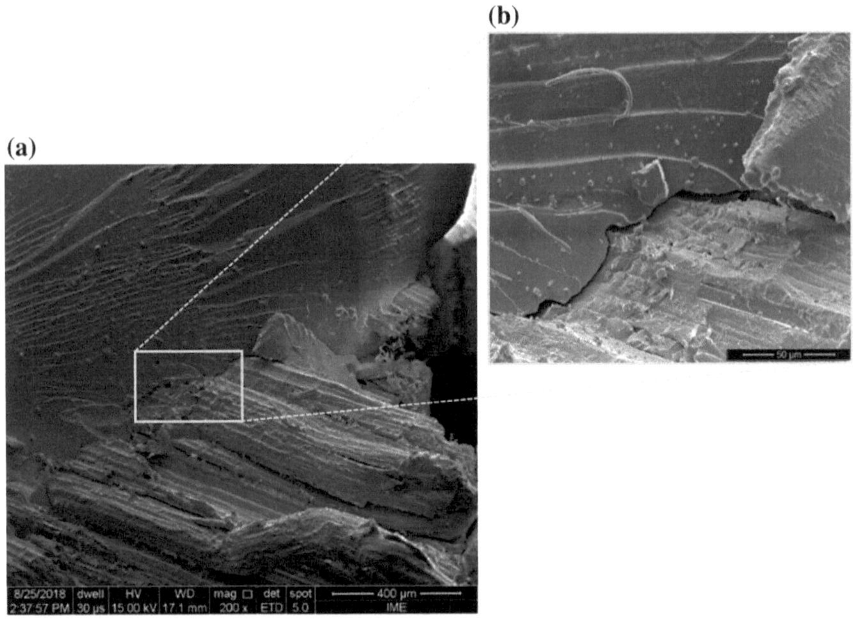

Fig. 3 Scanning electronic micrograph: **a** SEM of the interface between fique fibers and epoxy matrix (200×) **b** Magnification of interface in the fique fibers reinforced composite (1500×)

Summary and Conclusions

- Based on the analysis of the results obtained in this work, it was possible to evaluate the effect of the amount of reinforcement inserted in the epoxy matrix composites in terms of degree of crystallinity and morphological aspect of the fiber "*in natura*" as well as the fiber/epoxy interface.
- The fibers of the fique plant presented a semicrystalline characteristic and low degree of crystallinity when compared to other fibers.
- The SEM analysis was able to reveal the fiber morphology, which exhibit a porous surface. This porosity may be able to contribute to fiber/matrix interaction.

Acknowledgements The authors thank the Brazilian agencies CAPES and CNPq for the financial support.

References

1. Monteiro SN et al. (2018) Fique fabric: a promising reinforcement for polymer composites. Polymers 10(246):10
2. Nascimento LFC (2017) Characterization of the Epoxy-Malva fiber composite for use in multilayer ballistic shielding. Doctoral thesis on materials science. Military Engineering Institute. Rio de Janeiro
3. Benini KCCDC (2011) Development and characterization of reinforced polymeric composites with ligno-cellulosic fibers: hips/fiber from coco green shell and sugar cane bagasse. Faculty of Engineering of Campus of Guaratinguetá. Paulista State University. Mechanical Engineering in the field of Materials
4. Carvalho KCCD et al. (2010) Preparation and characterization of cane bag fibers modified with zirconium oxide nanoparticles. In: 19th Brazilian congress of materials engineering and science—CBECiMat. Campos do Jordão, SP, Brazil, pp 7726–7734
5. Rocha JG, Mulinari DR (2014) Mechanical characterization of LDPE composites reinforced with palm fibers. In: Ca-dernos UniFOA—Special edition of the professional master's course in materials, pp 45–53
6. El-Sabbagh A (2014) Effect of coupling agent on natural fibre in natural fibre/polypropylene composites on mechanical and thermal behaviour. Composites Part B 57:126–135
7. Nuthong W, Uawongsuwan P, Pivsa-Art W, Hamada H (2013) Impact property of flexible epoxy treated natural fiber reinforced PLA composites. Energy Procedia. 34:839–847
8. Ramesh M, Palanikumar K, Reddy KH (2013) Mechanical property evaluation of sisal–jute—glass fiber reinforced polyester composites. Composites Part B Eng 48:1–9
9. Arrakhiz FZ, El Achaby M, Malha M, Bensalah MO, Fassi-Fehri O, Bouhfid R, Benmoussa K, Qaiss A (2013) Mechanical and thermal properties of natural fibers reinforced polymer composites: doum/low density polyethylene. Mater Des 43:200–205
10. Segal L, Creely J, Martin AE Jr, Conrad CM (1959) An empirical method for estimating the degree of crystallinity of native cellulose using the X-ray diffractometer. Text Res J 29:786–794
11. Gañán P, Mondragon I (2002) Surface modification of fique fibers. effects on their physico-mechanical properties. Polym Composites 383–394
12. Hidalgo-Salazar MA, Correa JP (2018) Mechanical and thermal properties of biocomposites from nonwoven industrial fique fiber mats with epoxy resin and linear low density polyethylene. Results Phys 8:461–467
13. Hidalgo-Salazar MA, Mina JH, Herrera-Franco PJ (2013) The effect of interfacial adhesion on the creep behaviour of LDPE–Al–Fique composite materials. Composites Part B Eng 55: 345–351
14. Hoyos CG, Vázquez A (2012) Flexural properties loss of unidirectional epoxy/fique composites immersed in water and alkaline medium for construction application. Composites Part B Eng 43: 3120–3130
15. Hoyos CG, Alvarez VA, Rojo PG, Vázquez A (2012) Fique fibers: enhancement of the tensile strength of alkali treated fibers during tensile load application. Fibers Polym 13(5):632–640
16. Luna P, Mariño A, Lizarazo-Marriaga J, Beltrán O (2017) Dry etching plasma applied to fique fibers: influence on their mechanical properties and surface appearance. Proc Eng 200:141–147
17. Ovalle-Serrano SA, Blanco-Tirado C, Combariza MY (2018) Exploring the composition of raw and delignified Colombian fique fibers, tow and pulp. Cellulose 25(151)

Part V
Biomass in Armor Composites

Izod Impact Test on Epoxy Composites Reinforced with Mallow Fibers

Lucio Fabio Cassiano Nascimento, Sérgio Neves Monteiro, Ulisses Oliveira Costa and Luana Cristyne da Cruz Demosthenes

Abstract The mallow fibers (Urena lobata, linn) demonstrate great potential for the use as reinforcement in polymer matrix composites. In this work, epoxy composites with mallow fibers were produced in volumetric fractions of 0 and 10%, with the objective of investigating the average Izod impact energies. For the composites with 0 and 10% of mallow fibers, there was ineffective performance of the reinforcement, with the predominance of fragile fracture mechanisms, seen through SEM analysis. It was verified that there was an increase of the impact energy between the epoxy matrix and the composite with 10% of mallow fibers in 1697%.

Keywords Mallow fibers · Epoxy · Izod impact test

Introduction

Unlike synthetic fibers, natural lignocellulosic fibers (NLFs) do not have uniform properties, as they are microstructurally heterogeneous and have dimensional limitations. The same NLFs species may have their properties affected considerably depending on their source, plant quality and age, fiber diameter, aspect ratio, and preconditioning [1].

In the case of composites with continuous and aligned fibers, the mechanical characteristics depend on several factors, including the stress–strain behavior of the phases present, the volumetric fractions of these phases and the direction in which the stress or load is applied [2].

Mallow fiber (Urena lobata, linn), although widely used for the manufacture of blankets, carpets, paper money, handicrafts, and other various purposes, is not yet widely used as a reinforcing component in industrialized products. However, physical and mechanical properties such as tensile strength, Young's modulus, and impact

---

L. F. C. Nascimento (✉) · S. N. Monteiro · U. O. Costa (✉) · L. C. da Cruz Demosthenes
Materials science department, Military Institute of Engineering - IME, Rio de Janeiro – RJ, Brazil
e-mail: lucio_coppe@yahoo.com.br

U. O. Costa
e-mail: ulissesolie@gmail.com

S. Ikhmayies et al. (eds.), *Green Materials Engineering*, The Minerals, Metals & Materials Series, https://doi.org/10.1007/978-3-030-10383-5_16

Table 1 Physical and mechanical properties of mallow fibers. *Source* Adapted [3]

	Diameter (μm)	Length (mm)	Specific mass (kg/m^3)	Max absorption (%)	Tensile strength (MPa)	Elongation (%)	Young modulus (GPa)
Mallow	42.6	23.8	1374	377	160	5.2	17.4

Table 1 is from the dissertation of Jean Igor Margem reproduced with previous permission of the author, deceased in March 2018

Fig. 1 Mallow fibers as supplied by UENF

energy [3], [4], demonstrate great potential as reinforcement in polymer matrix composites (CNLFs). Some of these properties are shown in Table 1.

In this context, the possibility of using CNLFs in dynamic applications is highlighted. Therefore, it is essential to measure and adjust the Izod impact energy (J/m) in these composites.

The objective of this study is to manufacture epoxy composites reinforced with mallow fibers, in the volumetric fractions of 0 and 10% of fibers, in order to calculate the Izod impact energy absorbed by the CNLFs, showing the relation of this energy obtained with the volumetric fraction of mallow fibers. For this, the composites will be cut and the Izod test will be carried out under the conditions provided by ASTM D256, with the objective of obtaining the impact energy in J/m.

Materials and Methods

Mallow fibers supplied by the Universidade Estadual do Norte Fluminense (UENF) were used, which in turn were purchased from Companhia Têxtil de Castanhal. Figure 1 shows the mallow fibers as provided.

The polymer used as the matrix material of the composite board was commercially available epoxy resin, bisphenol A diglycidyl ether type (DGEBA), hardened with triethylene tetramine (TETA), using the stoichiometric ratio of 13 parts of hardener to 100 parts of resin. The resin company was Dow Chemical of Brazil, being supplied by the distributor RESINPOXY Ltda.

Preparation of Mallow Fibers

The fibers were cleaned, cut in the dimension of 150 mm, and dried in a greenhouse for 24 h, and then, they were used to make the composite plates.

Preparation of Epoxy-Mallow Fibers Composites

To make the composites, a metal matrix (Fig. 2) with an internal volume of 214.2 cm^3 (dimensions $15 \times 12 \times 1.19$ cm) was used. The plates were pressed in a hydraulic press with a maximum capacity of 30 tons. A load of 5 tons during was used 24 h to fabricate the composite bodies [4–7]. For the mallow fibers, the density of 1.40 g/cm^3 [8, 9] was used as initial reference and the value of 1.11 g/cm^3 [10] was used for the epoxy resin (DGEBA-TETA). The percentages of mallow used in the work were 0 and 10% v/v.

Fig. 2 Metal matrix for making composite plates

Fig. 3 Composite plates with percentages of fibers of 0 and 10%

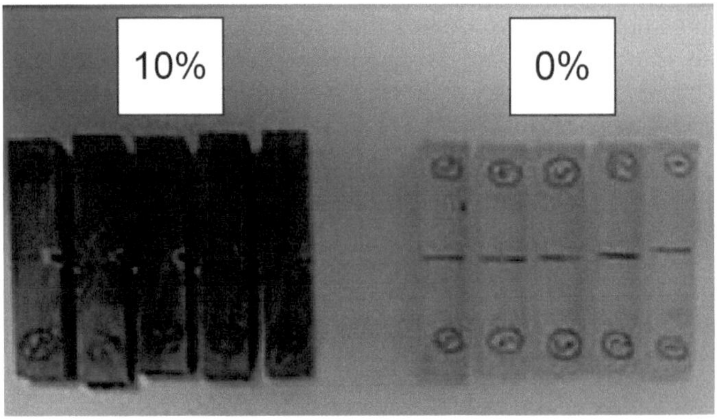

Fig. 4 Specimens with mallow fibers for Izod impact test

Preparation of Samples for Izod Impact Test

The Izod impact test specimens were produced from the cut of the composite strip saw blades in the dimensions provided for ASTM D256. The objective was to measure the fracture energy in Joules/meter (J/m). Five samples were produced for each percentage of fibers tested, with the dimensions of 62.5 × 12.7 × 10 mm. In Figs. 3 and 4, the composite plates and the samples prepared for the test are shown.

Table 2 Results the Izod impact test for epoxy matrix composites reinforced with continuous and aligned mallow fibers

Energy absorbed (J/m)—Izod test		
CPs	0%	10%
1	10.15	102.71
2	6.84	122.10
3	4.08	137.76
4	5.46	104.07
5	5.47	103.87
6	5.37	122.22
7	6.76	99.71
Average	6.30	113.21
Standard deviation	2.11	14.30

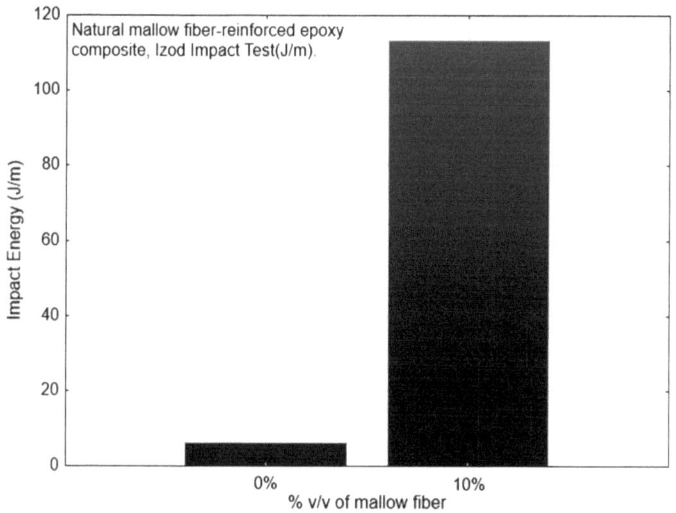

Fig. 5 Izod impact energy versus percentages of mallow fibers

Results and Discussion

From the results obtained and presented in Table 2 we can see a great jump in the Izod impact energy, from 0 to 10% v/v of fibers, about 16.97 times. The explanation for this change is associated with the transverse propagation of crack in the composite with 0% reinforcement, that is, as the fiber volume fraction increases, they become functioning as barriers that deviate the path of the cracks, and therefore, increases the system's impact energy [11].

Figure 5 shows the graph that relates the Izod impact energy to the percentage of mallow fibers. It is clearly seen that the impact energy increased with the addition of fibers [3, 12, 13].

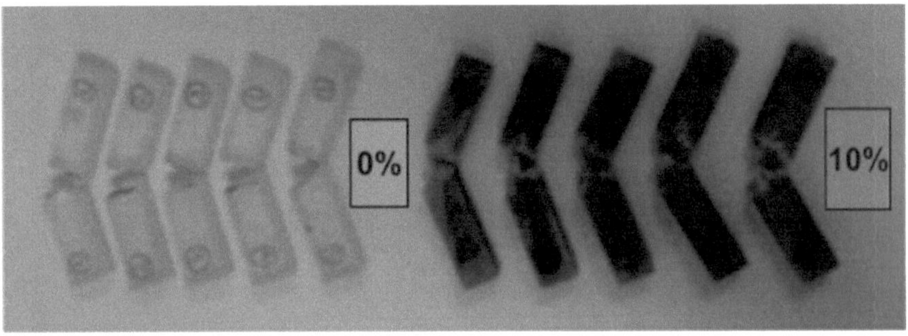

Fig. 6 Test specimens fractured completely after Izod impact test

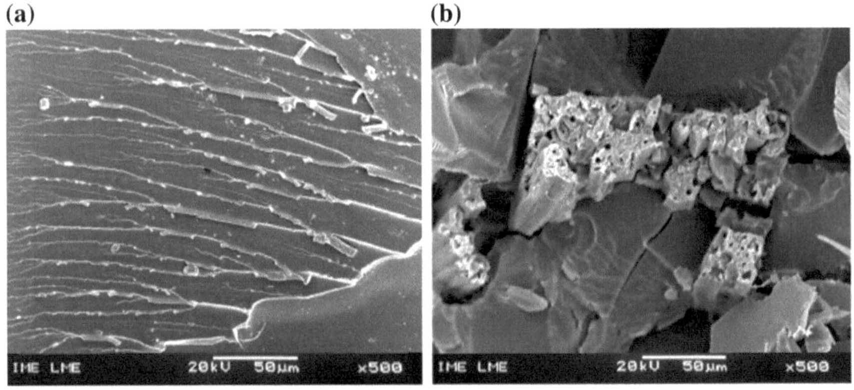

Fig. 7 Scanning electron microscopy of fracture surfaces of composites reinforced with mallow fibers after Izod impact test (500x). **a** 0% and **b** 10% v/v fibers

A similar result was also confirmed by other work involving impact energy in CLCFs [3, 11, 13]. Proving, therefore, the direct relation between the increase of the fiber volumetric fraction and the Izod impact energy.

Figures 6 and 7 confirm the tendency of fragile fracture in specimens of 0 and 10%, by the presence of "river mark" patterns (Fig. 7a). The crack propagates transversely and catastrophically in both, 100% epoxy composite and 10% fiber composite (Fig. 7b), showing that there was no effective reinforcement with this percentage of fibers [11, 13].

Conclusions

The Izod impact energy increases as a certain volumetric fraction of mallow fibers incorporated in the epoxy matrix is added. There was a predominance of fragile fracture in both cases. Therefore, this volumetric fraction of fibers was not sufficient to generate a change in the mechanism of fracture from ductile to brittle, showing that this percentage of fibers is relatively inefficient, although promoting in the composite material a mean Izod impact energy in the order of 113.21 J/m, higher than only epoxy matrix by approximately 16.97 times.

Acknowledgements The authors acknowledge the support to this investigation by the Brazilian agencies CNPq, CAPES, FAPERJ and UENF for supplying the mallow fibers.

References

1. Mohanty AK, Misra M, Hinrichsen G (2000) Biofibres, biodegradable polymers and bio-composites: an overview. Macromol Mater Eng 276(277):1–24
2. Callister WD, Rethwisch DG (2016) Science and engineering materials: an introduction, 9th edn. Rio De Janeiro: Ltc
3. Margem JI (2013) Study of the structural characteristics and properties of polymer composites reinforced with mallow fibers. PhD thesis (In Materials Engineering). Universidade Estadual Do Norte Fluminense, Rio De Janeiro
4. Luz FS (2014) Evaluation of ballistic behavior of multi-layered shielding reinforced epoxy composite with jute fiber. Post-Graduate course master's dissertation (In Materials Science). Military Institute of Engineering
5. Milanezi, TL, (2015) Evaluation of the synergic ballistic behavior of multi-layered shielding with the rami fiber. Post-graduate course master's dissertation (In Materials Science). Military Institute of Engineering
6. Braga FO (2015) Ballistic behavior of a multi-layered shield using polyester-curau composite as intermediate layer. Post-Graduate course master's dissertation (In Materials Science). Military Institute of Engineering
7. Agopyan V, Savastano Jr H (1997) Use of alternative materials based on vegetable fibers in civil construction: Brazilian experience. In: Ibero-American Seminar
8. Oliveira JTS (1998) Characterization of eucalyptus wood for civil construction. 429f. Doctorate thesis (In Civil Engineering)-Polytechnic School of The University Of São Paulo, São Paulo
9. Silva LC (2014) Ballistic behavior of the epoxy-curaua composite in multilayer shielding. Doctoral degree thesis (In Materials Science). Military Institute of Engineering, Rio De Janeiro
10. Gomes MA (2015) Mechanical properties of polymeric composites reinforced with pineapple leaf fibers (PALF). PhD thesis (In Materials Engineering). Universidade Estadual do Norte Fluminense, Rio de Janeiro
11. Bledzki AK, Gassan J (1999) Composites reinforced with cellulose based fibers. Prog Polym Sci 4:221–274
12. Candido VS (2014) Characterization and properties of polymeric composites reinforced with sugarcane bagasse fibers. PhD thesis (In Materials Science), Military Engineering Institute, Rio De Janeiro
13. D256 A (2005) Standard test method for determining the Izod pendulum impact resistance of plastics. American Society for Testing Materials International, Philadelphia, PA

Evaluation on the Design of Piassava Fiber Reinforcement Epoxy Matrix Composite for Ballistic Application

Fabio da Costa Garcia Filho, Luana Cristyne da Cruz Demosthenes, Michelle Souza Oliveira and Sergio Neves Monteiro

Abstract Since the last decade, some studies have been carried out regarding the use of natural lignocellulosic fibers (NLFs) as reinforcement in polymer matrix composites for ballistic application. The results obtained were very promising and could be even compared to Kevlar^{TM} when used as an intermediate layer in a multilayered armor system (MAS). Although some different NLFs had been tested, studies about the ideal design or configuration of the composite yet to be done. This work aims to evaluate four possible designs of an epoxy matrix composite reinforced with NLF, in this case piassava fibers, to be used as intermediate layer in a MAS. The performance of these composites as stand-alone ballistic armor was evaluated against ballistic impact of high energy 7.62 mm ammunition. The results were statistically analyzed using the analysis of variance (Anova) and Tukey's honest significant difference test (HSD). Furthermore, it was possible to determine the limit velocity (V_L), which is an important parameter for ballistic armors. The energy absorption was statistically the same for all four tested conditions but only three of them were able to keep it integrity after the ballistic impact.

Keywords Natural fiber · Ballistic application · Composites design

Introduction

The development on technologies of guns and ammunition results in more powerful and hazardous material; therefore, it has been an extensively effort to promote substantial improvements in the material used as ballistic armor. In a way to be considered as efficient materials for ballistic armor characteristics such as weightlessness

F. da Costa Garcia Filho (✉) · L. C. da Cruz Demosthenes · M. S. Oliveira · S. N. Monteiro
Department of Materials Science, Military Institute of Engineering—IME, Praça General Tibúrcio, 80, Praia Vermelha, Urca, Rio de Janeiro, RJ ZIP 22290-270, Brazil
e-mail: fabiogarciafilho@gmail.com

S. Ikhmayies et al. (eds.), *Green Materials Engineering*, The Minerals, Metals & Materials Series, https://doi.org/10.1007/978-3-030-10383-5_17

and resistance to perforation as well as high energy absorption are required [1, 2]. Synthetic fiber fabrics such as aramid (Kevlar™) and ultra-high molecular weight polyethylene (Dyneema®) are, currently, the mainly used material for ballistic armor application against high energy ammunitions, for instance, 7.62 and 5.56 mm [3, 4]. However, in the past five years, polymeric composites reinforced with natural lignocellulosic fibers (NLFs) have been reported as promising material when used in a multilayered armor system (MAS) [5–12].

Braga et al. [13] investigated the performance of natural curaua fiber-reinforced polyester composites as ballistic armor against high energy ammunition. Volume fraction of fiber reinforcement varying in the range of 0–30% was tested and the results revealed that the 30 vol% fiber composites were found to be the best alternative to Kevlar™ laminates.

Nascimento et al. [14] reported similar indentation after a perforation and backface signature test for MAS using epoxy composites reinforced with 30 vol% of mallow fibers when compared with those using Kevlar™ laminates as intermediate layer in the MAS. The comparable efficiency between the mallow reinforced composite and the Kevlar™ laminate was also supported by impedance matching calculation.

Assis et al. [15] showed the advantages on the use of jute non-woven mat-reinforced polyester matrix composite in multilayered armor. Fragments associated with the ballistic impact were successfully captured by the polyester/jute non-woven composite through mechanic incrustation, and this mechanism is similar to those reported by synthetic fibers used in ballistic applications. Furthermore, weight reduction as well as production cost reduction was considered as important advantages of this natural fiber-reinforced composite.

Although these promising ballistic results regarding the use of natural lignocellulosic fibers, most of the published papers considered only one-directional aligned fibers or two-dimensional fabrics reinforcing the composite. Therefore, further studies to compare the influence of the design of the composite are necessary. The fiber extracted from a Brazilian palm tree, *Attalea Funifera*, also known as piassava fiber is considered a promising candidate to reinforce polymeric matrices to obtain tougher and stiffer composites [16–18]. Thus, the objective of the present work is to evaluate the individual performance of these epoxy matrix composites reinforced with piassava fibers in four different designs in order to determinate the influence of the fiber disposition under high energy 7.62 mm ballistic impact.

Materials and Methods

Table 1 shows the materials tested in the present work as well as their designations. All fiber reinforcement volume fractions were kept as 30 vol%.

Table 1 Materials tested and their designations

Designation	Disposition
SRPF	Short and randomly scattered piassava fibers
LRPF	Long and randomly scattered piassava fibers
LAPF	Long and aligned piassava fibers, axial configuration
CPPF	90° Crossing and long piassava fibers, biaxial configuration

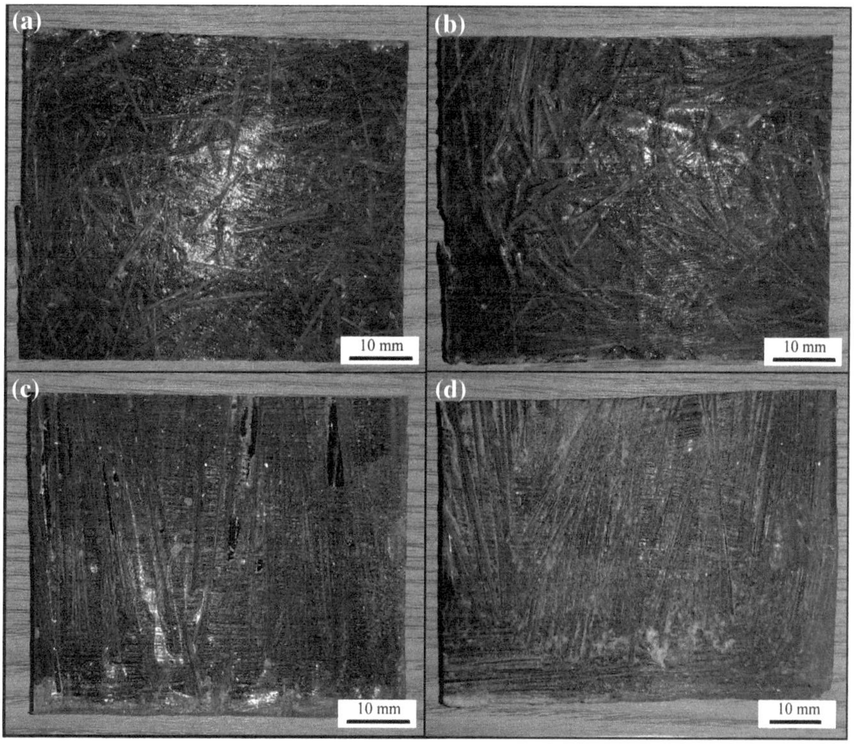

Fig. 1 Composite plates before ballistic test: **a** SRPF, **b** LRPF, **c** LAPF, and **d** CPPF

The piassava fibers were supplied by Vassouras Rossi, Brazil, in the form of bundles. The epoxy DGEBA/TETA, fabricated by Dow Chemical, was supplied by epoxy Fiber, Brazil. The composites plates, with dimensions of $60 \times 75 \times 10$ mm^3, were produced by the addition of the 30 vol% of piassava fiber with 70 vol% of epoxy resin into a metallic mold. A pressure of 3 MPa was applied and the curing process took place at room temperature. The criteria for determining are the fiber is long or short was based on a pull-out test [19]. If the length of the fiber is lower than 15 mm, it is considered as short and with length above this value the, fiber is long. Figure 1a–d presents the macroscopic aspect of the composites plates produced.

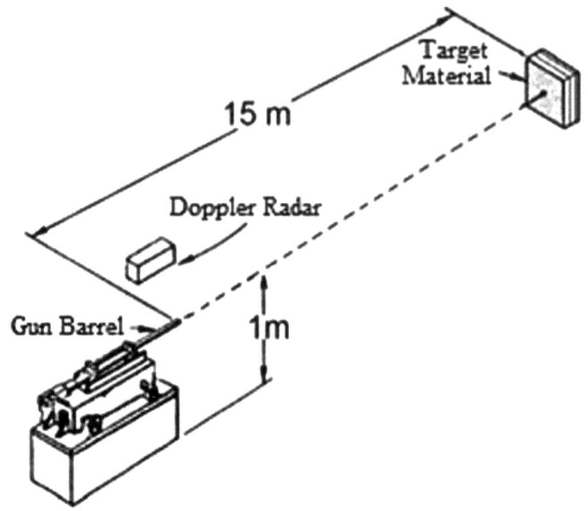

Fig. 2 Schematic illustration of the ballistic test

The ballistic tests were performed at the Brazilian Army Assessment Center (CAEx—*Centro de Avaliação do Exército*), in Rio de Janeiro, Brazil. A high-pressure instrumentation (HPI) gun barrel equipped with laser sight, model B290, was used as shooting device. The target material was placed 15 m from the gun barrel, and the bullet trajectory was described as perpendicular. The ammunition was a commercial 7.62 mm M1, full metal jacketed bullet of 9.7 g. The projectile velocity was measured before and after the ballistic impact with a model SL-520P Weibel Doppler radar provided with Windopp software to process the radar raw data. The schematic illustration of the ballistic test is shown in Fig. 2.

The kinetic energy variation of the projectile was related to the energy dissipated by the composite plate used as target. This variation of energy (E_D) can be calculated based on the bullet velocity before (V_b) and after (V_a) the impact as well as the bullet mass (m) shown in Eq. 1 [20]:

$$E_D = \frac{m}{2}(V_b^2 - V_a^2) \quad (1)$$

Furthermore, also based on the bullet measurement, the limit velocity (V_L) was determined for each condition investigated. After the calculation of the variation of energy dissipated, it is considered a situation where the velocity after the ballistic impact would be equal to zero. Therefore, in this scenario the projectile would be stopped by the target material. So the limit velocity (V_L) could be estimated by Eq. 2 [20]:

$$V_L = \sqrt{\frac{2E_D}{m}} \quad (2)$$

The results were statistically treated using the analysis of variance (Anova) [21]. This is a powerful analysis that verifies with a high degree of confidence is the treatment, in this case the design of the composite, indeed, influenced in the results obtained. This analysis is based on statistical parameters such as degrees of freedom (DF), sum of squares (SS), mean squares (MS), the Snedecor F and Fc, calculated and critical. After that, Tukey's test, a further statistical test was performed [21]. Such test compares the mean values two by two, in order to quantify if there is a significant difference among the tested conditions, this test is also called honestly significant difference (HSD) and is calculated by Eq. 3:

$$HSD = q\sqrt{\frac{EMS}{r}} \quad (3)$$

Results and Discussion

Figure 3 exhibits typical Doppler radar data. In Fig. 3a, it is illustrated the reflected raw data, while in Fig. 3b its corresponding Windopp fitted radial velocity curve is presented. Analyzing the fitted curve in Fig. 3b, it is possible to verify the projectile with decreasing velocity exits the gun barrel with 848 m/s and hits the composite plate at $Vb = 833$ m/s. The projectile impact with the target is associated with the vertical line exhibit at about 0.018 s. One may verify that the projectile exits from the target with a residual velocity $Va = 775$ m/s. The shape of the single curve with continuously decreasing velocity indicates that the bullet was able to be kept intact after the ballistic impact. As aforementioned, the measurement of these velocity values allows the calculation of the energy dissipated by the target material.

Table 2 presents the mean value of energy dissipated by all conditions tested as well as the standard deviation.

Considering the results obtained, it seems that the energy dissipation was the same for every each condition. Yet, the statistical analysis is necessary to validate this data. Table 3 presents the Anova parameters. Comparing the F calculated with the Fc (tabled), it is possible to verify that F<Fc. Therefore, it is possible to affirm with 95% of confidence that the energy dissipation calculated is the same and, indeed, the target material does not influence in the result.

Nevertheless, the honestly significant difference test was performed in order to compare the mean values quantitatively two by two. Considering 4 treatments and 12 degrees of freedom for the error, the tabled value for q is 4.2. Thus, the HSD calculated by Eq. 3 was 39.4. Table 4 summarizes the results of Tukey's analysis.

Table 2 Energy dissipated by all conditions, mean value, and standard deviation

	SRPF	LRPF	LAPF	CPPF
Energy dissipated (J)	198±24	199±16	204±12	188±21

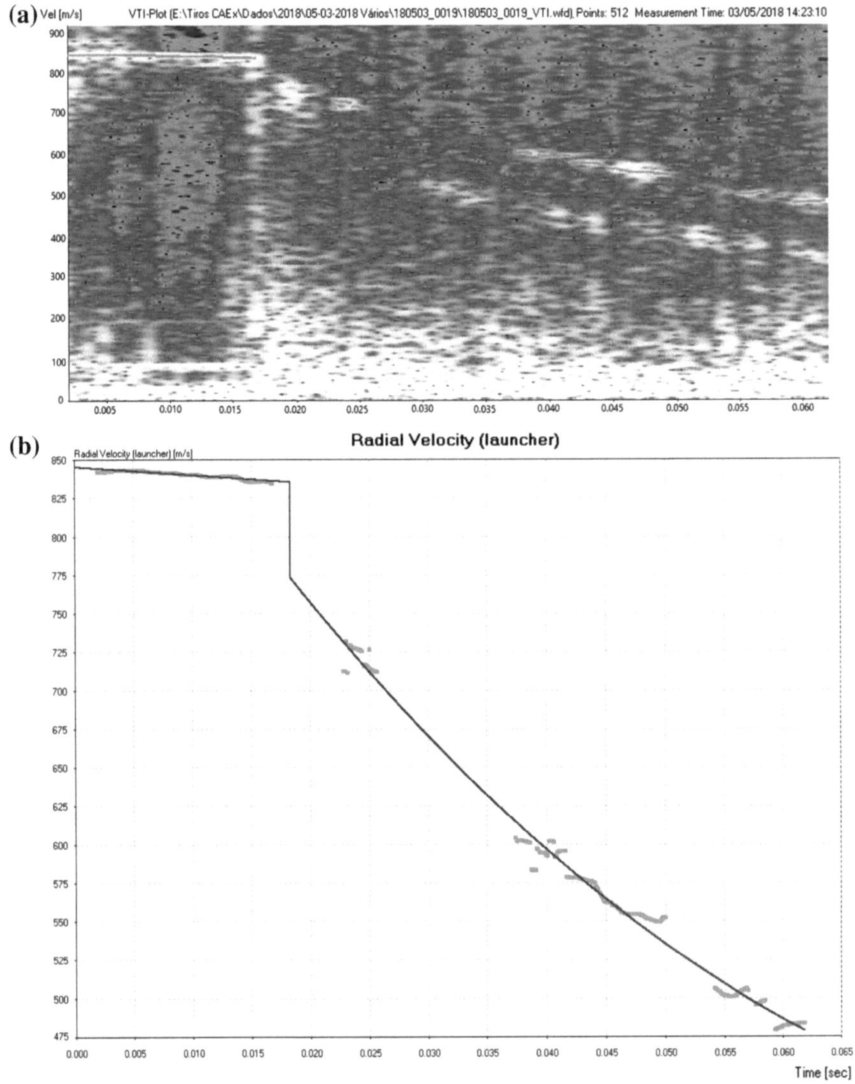

Fig. 3 Doppler radar raw data (**a**) and treated data (**b**)

Table 3 Analysis of variance (Anova) parameters

	DF	SS	MS	F	Fc
Treatments	3	537.7	179.2	0.51	3.49
Error	12	4219.3	351.6		
Total	15	4756.9			

Table 4 Tukey's honestly significant difference for the energy dissipation of the composites plates

Condition	SRPF	LRPF	LAPF	CPPF
SRPF	0	1	6	10
LRPF	1	0	5	11
LAPF	6	5	0	16
CPPF	10	11	16	0

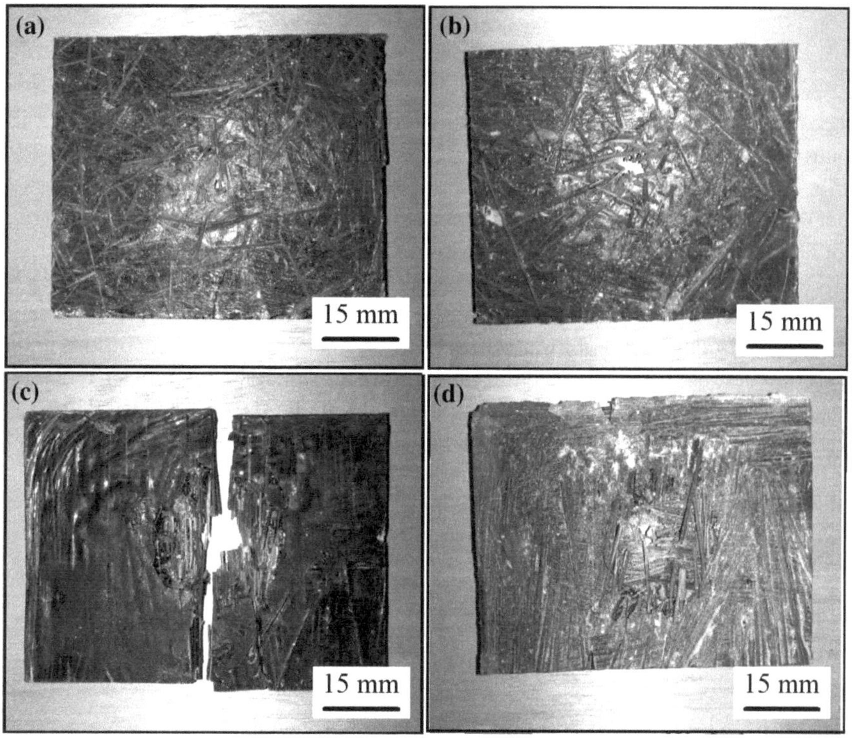

Fig. 4 Macroscopic aspect of the targets after the ballistic impact: **a** SRPF, **b** LRPF, **c** LAPF and **e** CPPF

These results are very interesting and statistically prove that none of the tested fiber disposition in the polymer matrix composite had a significantly impact in the energy dissipation from the ballistic test. On the other hand, when considered the macroscopic aspect of the composite plate after the ballistic impact, it is possible to notice clear difference, as shown in Fig. 4a–d.

One may notice that for the LAPF condition, Fig. 4c, the composite presented a critical fracture that tends to occur in the same direction of the fibers disposition. Some other researchers also reported similar behavior and justified this pattern of failure as function of the mismatched nature between the fibers and the polymeric matrix [22, 23]. It is important to realize that all other three conditions were able to

Table 5 Ballistic parameter of limit velocity

Material	V_L (m/s)	Reference
SRPF	202±17	[a]PW
LRPF	202±12	[a]PW
LAPF	205±9	[a]PW
CPPF	197±15	[a]PW
Kevlar™	212±23	[13]

[a]Present work

keep their integrity after the ballistic test. The disposition of the fibers randomly by the matrix SRPF (Fig. 4a) and LRPF (Fig. 4b) as well as the biaxial configuration CPPF (Fig. 4d) provides reinforcement in more than one direction. This analysis showed that the disposition of the reinforcement fibers aligned in an axial configuration may not be suitable for ballistic armor applications.

Finally, the limit velocity (V_L) was calculated for all conditions using Eq. 2. Table 5 presents these results as well as the value of this ballistic parameter for other materials. Similar values of limit velocity were found for the tested conditions in this present work, as expected, once the estimation of this ballistic parameter was made considering the energy dissipated by the material, which was proven to be statistically the same. One may verify that these parameters are in the same order of magnitude of other materials such as Kevlar™, which is considered efficient material against high energy ammunition menace. This may qualify these epoxy matrix-reinforced piassava fiber composites as promising materials for ballistic armor application.

Conclusions

- Four different dispositions of piassava fiber reinforcement in epoxy matrix composite were analyzed against high energy 7.62 mm ammunition. Conditions of axial and biaxial reinforcement as well as short and long scattered fibers were considered.
- The energy dissipated by the composites plate varied in the range from 188 to 204 J. Furthermore, it was statistically proved by the analysis of variance and honestly significant difference that the results were the same for all conditions investigated.
- Although the energy dissipated by the composites plates were the same, the physical integrity was also considered as restricting to the use of these materials as ballistic armor. The conditions where the displacement of the fibers provide a reinforcement in more the one direction, i.e., SRPF and LRPF (randomly scattered in the matrix) as well as CPPF (90° cross-ply) were able to keep their integrity after the ballistic impact. On the other hand, the LAPF presented critical fracture trending to occur in the same direction of fibers alignment.

- The ballistic parameter limit velocity was calculated and shown to be the same as others materials, such as Kevlar.

Acknowledgements The authors thank the support of this investigation by the Brazilian agencies: CNPq, FAPERJ and CAPES.

References

1. Wang L, Kanesalingam S, Nayak R, Padhye R (2014) Recent trends in ballistic protection. TLIST 3:37–47
2. Medvedovski E (2010) Ballistic performance of armour ceramics: influence of design and structure. Part 1. Ceram Int 36:2103–2115
3. Akella K, Naik NK (2015) Composite armor—a review. J Indian Inst Sci 95(3):297–312
4. Monteiro SN, Lima EP Jr, Louro LHL, Silva LC, Drelich JW (2014) Unlocking function of aramid fibers in multilayered ballistic armor. Metall Mater Trans A 46A:37–40
5. Benzait Z, Trabzon L (2018) A review of recent research on materials used in polymer-matrix composites for body armor application. J Compos Mater 0(0):1–23. https://doi.org/10.1177/0021998318764002
6. Pickering KL, Aruan Efendy MG, Le TM (2016) A review of recent developments in natural fibre composites and their mechanical performance. Composites: Part A 83:98–112
7. Braga FO, Bolzan LT, Luz FS, Lopes PHLM, Lima EP Jr, Monteiro SN (2017) High energy ballistic and fracture comparison between multilayered armor systems using non-woven curaua fabric composites and aramid laminates. J Mater Res Technol 6(4):417–422
8. Luz FS, Lima EP Jr, Louro LHL, Monteiro SN (2015) Ballistic test of multilayered armor with intermediate epoxy composite reinforced with jute fabric. Mater Res 18:170–177
9. Monteiro SN, Assis FS, Ferreira C, Simonassi NT, Weber RP, Oliveira MS, Colorado H, Pereira AC (2018) Fique fabric: a promising reinforcement for polymer composites. Polymers 10:246–245
10. Luz FS, Monteiro SN, Lima ES, Lima EP Jr (2017) Ballistic application of coir fiber reinforced epoxy composite in multilayered armor. Mater Res Ibero Am J Mater 20:23–28
11. Monteiro SN, Braga FO, Lima EP Jr, Louro LHL, Drelich JW (2017) Promising curaua fiber-reinforced polyester composite for high-impact ballistic multilayered armor. Polym Eng Sci 57:947–954
12. Nascimento LFC, Louro LHL, Monteiro SN, Gomes AV, Marcal RLSB, Lima EP Jr, Margem JI (2017) Ballistic performance of mallow and jute natural fabrics reinforced epoxy composites in multilayered armor. Mater Res Ibero Am J Mater 20:399–403
13. Braga FO, Bolzan LT, Lima Jr EP, Monteiro SN (2017) Performance of natural curaua fiber-reinforced polyester composites under 7.62 mm bullet impact as a stand-alone ballistic armor. J Mater Res Technol 6(4):323–328
14. Nascimento LFC, Louro LHL, Monteiro SN, Lima EP Jr, Luz FS (2017) Mallow fiber-reinforced epoxy composites in multilayered armor for personal ballistic protection. JOM 69:2052–2056
15. Assis FS, Pereira AC, Garcia Filho FC, Lima Jr EP, Monteiro SN, Weber RP (2018) Performance of jute non-woven mat reinforced polyester matrix composite in multilayered armor. J Mater Res Technol-JMR&T. DOI:https://doi.org/10.1016/j.jmrt.2018.05.026
16. Monteiro SN, Lopes FPD, Ferreira AS, Nascimento DCO (2009) Natural fiber polymer matrix composites: cheaper, tougher and environmentally friendly. JOM 61:17–22. https://doi.org/10.1007/s11837-009-0004-z
17. Nascimento DCO, Ferreira AS, Monteiro SN, Aquino RCMP, Satyanarayana GK (2012) Studies on the characterization of piassava fibers and their epoxy composites. Composites: Part A 43:353–362

18. Satyanarayana KG, Guimarães JL, Wypych F (2007) Studies on lignocellulosic fiber of Brazil. Part I: Source, production, morphology, properties and applications. Composites: Part A 38(1):694–709
19. Aquino RCMP, Monteiro SN, D'Almeida JRM (2003) Evaluation of the critical fiber length of piassava (Attalea funifera) fibers using the pullout test. J Mater Sci Lett 22:1495–1497
20. Morye SS, Hine PJ, Duckett RA, Carr DJ, Ward IM (2000) Modelling of the energy absorption by polymer composites upon ballistic impact. Compos Sci Technol 60:2631–2642
21. Vieira S (2006) Análise de Variância (Anova). Ed. Atlas, São Paulo
22. Pereira AC, Monteiro SN, Assis FS, Margem FM, Luz FS, Braga FO (2017) Charpy impact tenacity of epoxy matrix composites reinforced with aligned jute fibers. J Mater Res Technol-JMR&T 6(4):312–316
23. Oliveira CG, Margem FM, Monteiro SN, Lopes FPD (2017) Comparison between tensile behavior of epoxy and polyester matrix composites reinforced with eucalyptus fibers. J Mater Res Technol-JMR&T 6(4):406–410

Ballistic Test of Multilayered Armor with Intermediate Polyester Composite Reinforced with Fique Fabric

Artur Camposo Pereira, Foluke Salgado de Assis, Fabio da Costa Garcia Filho, Luana Cristyne da Cruz Demosthenes, Henry Alonso Colorado Lopera and Sergio Neves Monteiro

Abstract Multilayered armor systems (MASs) with a front ceramic followed by synthetic fabric are currently used against high-velocity ammunition. In these armors, the front layer, which shatters the ceramic and spalls the bullet, is followed by an intermediate layer, usually plies of Kevlar® or Dyneema®. In the present work, the intermediate synthetic fabric layer was replaced by an equal thickness layer of 30 vol.% fique fabric-reinforced polyester composite. The fique fabric is made of relatively strong natural fiber with a reinforcement potential. Ballistic impact test with 7.62 caliber ammunition revealed that both the plain polyester and the fique fabric composite have a relatively similar performance as that of the Kevlar®. The MAS with fique fabric component attended the NIJ standard for body protection. The energy dissipation mechanisms of fique fabric composite were analyzed by scanning electron microscopy. This latter is the same mechanism recently disclosed for aramid fabric.

Keywords Natural fiber composites · Fique fabrics · Multilayered armor · Ballistic test

Introduction

Ballistic protection of personnel, equipment and vehicles is now of surging importance due to armed urban conflicts and regional wars involving increasing firepower. The use of high velocity, impact, and power ammunition, such as class III [1], 7.62 × 51 mm (7.62 mm), constitutes a major personal threat. In this case, single layer shield-

---

A. C. Pereira (✉) · F. S. de Assis · F. da Costa Garcia Filho · L. C. da Cruz Demosthenes · S. N. Monteiro
Instituto Militar de Engenharia, IME, Praça Gen. Tibúrcio, N°80, Rio de Janeiro - RJ, Urca 22290-270, Brazil
e-mail: arturcamposomv@hotmail.com

H. A. C. Lopera
University of Antioquia, Mechanical Engineering, Bloque 20 of 437, Calle 67# 53-108, Medellin, Colombia

S. Ikhmayies et al. (eds.), *Green Materials Engineering*, The Minerals, Metals & Materials Series, https://doi.org/10.1007/978-3-030-10383-5_18

Table 1 Mechanical properties of the fique fibers [10, 13]

Tensile strength (MPa)	237 ± 51
Strain at break (%)	6.02 ± 0.69
Tensile modulus (GPa)	8.01 ± 1.47
Average diameter (mm)	0.16

ing, such as those made with plain Kevlar® (aramid fibers), would require a large thickness, which interferes with the user's mobility. For protection, the multilayer armor system (MAS), normally composed of three different materials, is intended for class III ballistic protection associated with a relatively smaller thickness allowing greater user mobility. As a front layer, a ceramic material is chosen due to its high compressive strength and hardness. Indeed, with a hard and brittle front material, the MAS is an effective solution [2–5]. As an intermediate layer, aramid fabric laminates, such as Kevlar® and Twaron® [2–6], as well as ultra-high molecular weight polyethylene, such as Dyneema® and Spectra® [7, 8], have been commonly used. This layer has the purpose of preventing the penetration of the fragments of both the ceramic and the projectile, resulting from the projectile impact [2–8]. Finally, as a third layer, the MAS may have a light and ductile metallic material, such as an aluminum alloy, to further dissipate the bullet impact shock wave energy [9].

In previous works, leftover fique fibers (*Furcraea andina*) were shown to be a suitable reinforcement element for polymer matrix composites [10–13]. The basic description of the fibers' mechanical properties and morphological characteristics is presented in Table 1.

The application of fique fabric polyester composite has been shown as promising MAS second layer [13]. In fact, it might contribute to prevent the lethal trauma in the human body, based on the ballistic behavior of the natural fiber composite [14], which takes into consideration the safety depth of indentation made in a clay witness positioned behind the MAS. According to the standard [1], the indentation, also known as backface signature, should be smaller than 1.73 inch (44 mm), in order to be considered efficient.

Therefore, the objective of this work was to evaluate the ballistic behavior of 10, 20, and 30 vol.% fique fabric-reinforced polyester composites as a MAS second layer subjected to 7.62 mm ballistic impact. It also investigated the energy dissipation mechanisms of fique fabric composite by scanning electron microscopy.

Experimental Procedure

The MAS investigated in the present work was composed of a front hexagon-shaped (30 mm in edge and 10 mm in thickness) Al_2O_3–4%Nb_2O_5 ceramic tile. A rectangular plate (120 × 150 × 10 mm) was used as second layer of polyester composites reinforced with fique fabric. A back layer of 5052 H34 aluminum alloy, also as rectangular plate (120 × 150 × 5 mm), finished the MAS. The layers were bonded together using a polyurethane adhesive.

The ceramic powders, Al_2O_3 (acquired from Treibacher Schleifmittel) and Nb_2O_5 (acquired from the Companhia Brasileira de Metalurgia e Mineração-CBMM), were mixed in a 4 wt% Nb_2O_5 proportion. The mixture was then ball milled in aqueous suspension, dried at 60 °C for 48 h, sifted until the 0.355 screen, cold pressed at 30 MPa, and sintered at 1400 °C for 3 h.

For the composite production, an orthophthalic polyester resin hardened with 0.5 wt% of methyl ethyl ketone, both supplied by the Resinpoxy, Brazil, was used as polymeric matrix. The fique fabrics were supplied by one of the co-authors (HACL) from Colombia. The fabrics were manually cleaned, cut, and dried at 60 °C for 24 h for the production of composites. Continuous and aligned fiber composite plates (120 × 150 × 10 mm) were prepared by compression molding, at room temperature (25 °C), in the volumetric fractions of 10, 20, and 30 vol.%. Plain polyester plates were similarly prepared. The fabrics were carefully positioned in the mold. After adding the resin with hardener, the mixture was kept under a pressure of 5 MPa for 24 h. Third MAS layer of the 5052 H34 aluminum alloy, with same 120 × 150 × 10 mm dimensions, was cut from sheets supplied by Metalak Metais, Brazil.

Ballistic tests were performed with NATO 7.62 × 51 mm military ammunition, following the NIJ procedures [1]. This means that the MAS target was positioned in front of a Roma Plastilina type of clay witness that simulates the consistency of the human body, 15 meters away from the shooting device. The experimental arrangement is shown in Fig. 1a and a real MAS mounted for the test in Fig. 1b. The shooting device was a model B290 gun barrel with laser sight (Fig. 1c), produced by HPI-High Pressure Instrumentation, available at the Brazilian Army Assessment Center (CAEx), located in the Marambaia peninsula, Rio de Janeiro, Brazil.

The bullet velocity was measured by both optical barriers and a model SL-520P Weibel doppler radar. The velocity was kept in the 847 ± 9.1 m/s range, measured 2.5 m before the target, as specified by the NIJ standard [1]. The indentation imprinted by the armor in the clay witness after the impact (backface signature) was measured by laser sensor and taken as a measure of the ballistic performance. The standard specifies that the mean indentation be no deeper than 1.73 inch (44 mm) [1]. Figure 1d depicts a measurement of the indentation after the impact. Six samples of each group were tested, and the data were statistically treated by the Weibull analysis, which gave information about the indentation distribution of probability [15]. The Weibull's density of the probability function is given by Eq. 1.

$$F(x) = 1 - \exp\left[-\left(\frac{x}{\theta}\right)^{\beta}\right] \tag{1}$$

Where x is the value assumed by the random variable; θ is the scale parameter; and β is the shape parameter, also known as Weibull modulus.

In order to identify the failure modes of the MAS, a microscopic evaluation of the fragments was carried out using the scanning electron microscope (FEI Quanta FEG 250), using secondary electrons and 20 kV for the acceleration voltage.

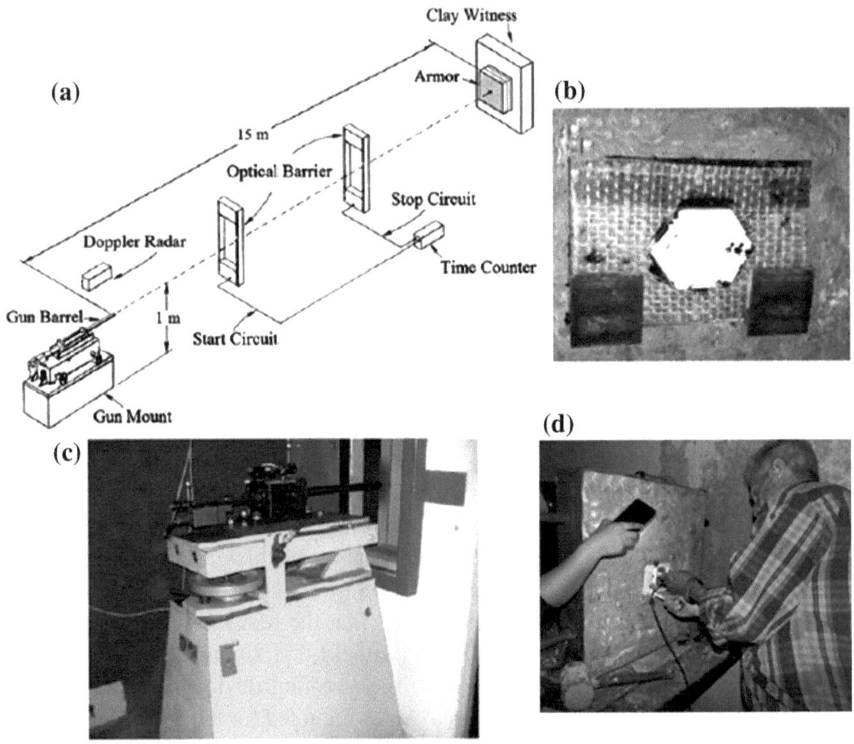

Fig. 1 Ballistic test: **a** schematic diagram with the experimental arrangement; **b** MAS positioned in front of the clay witness; **c** gun barrel with laser sight; **d** measurement of the indentation after the impact

Results and Discussion

Figure 2 shows aspect of MAS targets with reinforced polyester composite with 10, 20, and 30% fique fabric after the projectile impact. As noted in this figure, MASs with polyester composite reinforced with fique fabric did not exhibit perforation of the projectile, that is, the armor systems absorbed their kinetic energy. In all cases, there was total destruction of the ceramic material.

By means of the penetration in the clay witness, Table 2 shows the values obtained for the indentation depth associated with the different MAS ballistic tested.

Table 3 presents corresponding calculated Weibull parameters for the 10, 20, and 30% composites MAS. The high correlation coefficient (R^2) values indicate that just a set of failure mechanisms are occurring in both MAS, and this does not change in the observed indentation ranges. In other words, the variability of depth of indentation values within the groups was purely statistical. Indeed, the distribution of probabilities follows a Weibull distribution with high accuracy.

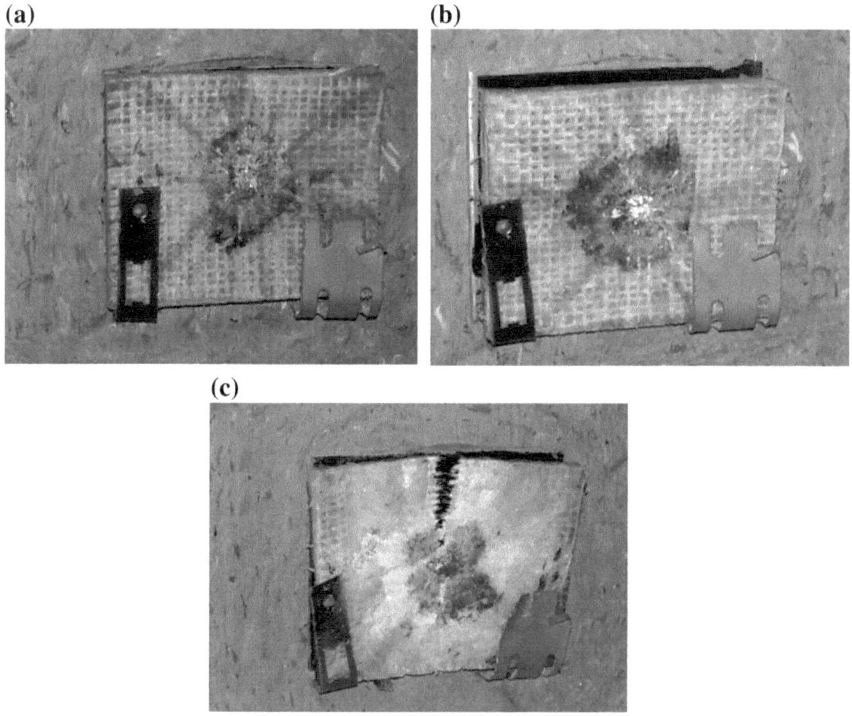

Fig. 2 View of the MAS target after the ballistic test, with second layer of **a** 10 vol.% fique fabric polyester composite; **c** 20 vol.% fique fabric polyester composite; **d** 30 vol.% fique fabric polyester composite

Table 2 Average depth of indentation in the clay witness backing different multilayered armors

Intermediate material layer	Average depth of indentation (mm)	Ref.
Polyester-30 vol.% fique fabric	20 ± 2	PW[a]
Polyester-20 vol.% fique fabric	17 ± 3	[13]
Polyester-10 vol.% fique fabric	16 ± 3	[13]
Aramid fabric laminates	23 ± 3	[15]

[a]Present Work

Table 3 Weibull parameters for ballistic tests with MAS having different second layers

Intermediate material layer	Weibull modulus (β)	Scale parameter (mm) (θ)	Correlation coefficient (R^2)
30 vol.%	13.61	17.18	0.98
20 vol.%	5.26	18.99	0.83
10 vol.%	9.84	20.84	0.86

Fig. 3 SEM of the fracture surface of the composite covered with ceramic fragments

Figure 3 illustrates SEM micrographs of the fracture aspects typical of the second layer for improving the MAS performance. These are mainly associated with the ability of collecting the fragments from both the projectile and the front ceramic layer, which is accomplished by mechanisms of mechanical incrustation, as well as attraction due to the presence of Van der Waals forces and short-living surface static charges.

Conclusions

Multilayered armor systems (MAS) using 10, 20, and 30 vol. % fique fabric-reinforced polyester were considered equally efficient for the protection against 7.62 mm caliber ammunition, based on the NIJ 0101.04 criteria related to the depth of indentation in the clay witness (single-hit protection).

The fracture mechanisms of the composites identified the importance of macroscopic delamination and microscopic capture of ceramic and projectile fragments.

Acknowledgements The authors thank the support for this investigation by the Brazilian agencies: CNPq, CAPES, and FAPERJ. The authors also acknowledge the permission to the use of the tensile equipment of COPPE/UFRJ.

References

1. National Institute of Justice NIJ Standard 0101.04 (2000) Ballistic Resistance of Personal Body Armor, US Depart. of Justice, Washington DC, September 2000
2. Li R, Fan Q, Gao R, Huo L, Wang F, Wang Y (2014) Effects of dynamic mechanical properties on the ballistic performance of new near-β titanium alloy Ti684. Mater Des 62:233–240
3. Tasdemirci A, Tunusoglu G, Guden M (2012) The effect of the interlayer on the ballistic performance of ceramic/composite armors: experimental and numerical study. Int J Impact Eng 44:1–9
4. Jacobs MJN, Van Dingenen JLJ (2001) Ballistic protection mechanisms in personal armor. J Mater Sci 36(13):3137–3142
5. Abrate S (1998) Ballistic impact on composite structures, 1st edn. Cambridge University Press, Cambridge, pp 215–220
6. Lee YS, Wetzel ED, Wagner NJ (2003) The ballistic impact characteristic of Kevlar® woven fabrics impregnated with a colloidal shear thickening fluid. J Mater Sci 38(13):2825–2833
7. Morye SS, Hine PJ, Duckett RA, Carr DJ, Ward IM (2000) Modeling of the energy absorption by polymer composites upon ballistic impact. Compos. Sci. Technol. 60(14):2631–2642
8. Lee BL, Song JW, Ward JE (1994) Failure of Spectra® polyethylene fiber-reinforced composites under ballistic impact loading. Compos. Mater. 28(13):1202–1226
9. Medvedovski E (2010) Ballistic performance of armor ceramics: influence of design and structure. Ceram Int 36(7):2103–2127
10. Gañán P, Mondragon I (2002) Surface modification of fique fibers. Effects on their physico-mechanical properties. Polym Compos 383–394
11. Teles MCA et al (2015) Fique fiber tensile elastic modulus dependence with diameter using the weibull statistical analysis. Mater Res 18:193–199
12. Netto PA et al (2016) Correlation between the density and the diameter of fique dibers. Mater Sci Forum 869:377–383
13. Monteiro SN et al (2018) Fique fabric: a promising reinforcement for polymer composites. Polymers 10:246–10
14. Rohen LA, Margem FM, Monteiro SN, Vieira CMF, Araújo BM, Lima ES (2015) Ballistic efficiency of an individual epoxy composite reinforced with sisal fibers in multilayered armor. Mater Res 18:55–62
15. Monteiro SN, Louro LHL, Trindade W, Elias CN, Ferreira CL, Lima ES et al (2015) Natural curaua fiber-reinforced composites in multilayered armor. Metall Mater Trans A 46(10):4567–4577

Ballistic Tests of Epoxy Matrix Composites Reinforced with Arapaima Fish Scales

Luís Carlos da Silva, Michelle Souza Oliveira, Luana Cristyne da Cruz Demosthenes, Wendell Bruno Almeida Bezerra and Sérgio Neves Monteiro

Abstract Arapaima fish, also known as pirarucu in Amazonian region, has strong and hard scales for protection against predators. The objective of this work is to investigate the ballistic efficiency of epoxy matrix composites reinforced with arapaima fish scales as second layer in multilayer armor. Multilayered armor systems have been widely used as ballistic armor, due to the contribution of the mechanical properties each layer of the system make has to absorb the kinetic energy of a projectile. The first layer generates ceramic fragments after the projectile impact. The second layer formed by the composite under study retains the fragments of the projectile and the ceramic while the last layer made of an aluminum alloy absorbs the residual energy by means of plastic deformation of the aluminum. The tests were carried out on three reinforced composites with 30% arapaima scales, in which a penetrating into the target was measured at the deformation in a clay witness. A 7.62 mm ammunition was used. The results indicate that epoxy matrix composites reinforced with arapaima fish scales display a ballistic performance comparable to Kevlar and natural fiber composites as MAS second layer.

Keywords Ballistics test · Arapaima fish scale · Epoxy composite · Weibull analysis

Introduction

Some bio materials are known for their toughness and flexibility. Fish scales have already been studied for a variety of applications [1]. Fish scales have drawn considerable attention due to their flexibility, their laminated composite structure, and their specific strength and toughness. These qualities have evolved to meet specific requirements of each fish according to its habitat and predatores [2]. Arapaima gigas

L. C. da Silva (✉) · M. S. Oliveira · L. C. da Cruz Demosthenes · W. B. A. Bezerra · S. N. Monteiro
Department of Materials Science, Military Institute of Engineering, IME, Praça General Tibúrcio 80, Urca, Rio de Janeiro 22290-270, Brazil
e-mail: lcarloseng@gmail.com

S. Ikhmayies et al. (eds.), *Green Materials Engineering*, The Minerals, Metals & Materials Series, https://doi.org/10.1007/978-3-030-10383-5_19

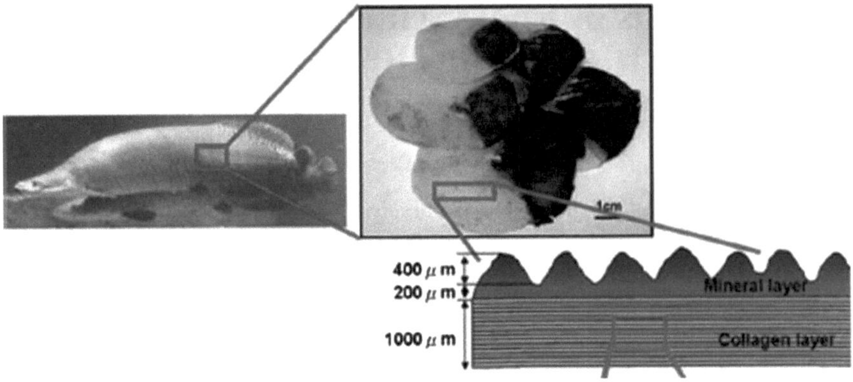

Fig. 1 Arapaima fish and details of scales. With permission from [6]

is one of the largest freshwater fish in the world, reaching a length of about 2–2.5 m and a mass over 150 kg. The fish inhabits the Amazon River Basin in South America. Interestingly, it lives in harmony with the piranha, a fish known for it voraciousness and sharp teeth. It has been proposed that the scales of Arapaimas serve as an armor-like protection against the sharp piranha teeth [3].Arapaima Scales are the skeletal elements that cover and protect the skin the animal [4]. These scales have a laminate composite structure composed of an external mineralized layer and internal lamellae with thickness of 50–60 μm each and composed of collagen fibers with ~1 μm diameter. The alignment of collagen fibers is consistent in each individual layer but varies from layer to layer, forming a nonorthogonal plywood type of structure, known as Bouligand stacking [3]. three scales overlapping essentially multiplies the puncture resistance by three. Friction between the scales is negligible, and therefore does not generate additional resistance to deformation or puncture, regardless of the arrangement of the scales. Indentation tests at different scales (macro-, micro-, and nano-) have also been used to assess the mechanical properties of fish scales. Lin et al. have reported an average hardness value in the internal layers of A. gigas scales of 200 MPa, increasing to about 550 MPa in the external layer. Indentation tests carried out by Torres et al. showed that the hardness of the cross section of A. gigas scale follows a cyclic sawtooth shaped pattern. The higher hardness values correspond to the laminates where fibres are perpendicular to the indentation plane. Bruet et al. found that the scales from Polypterus senegalus have hardness values that increase with distance from the inner surface to the outer surface, from 0.54 to 4.5 GPa [5]. The understanding of such flexible dermal armor is important as it may provide a basis for new synthetic, yet bioinspired, armor materials [6]. Figure 1 illustrates na arapaima fish and details of its scales.

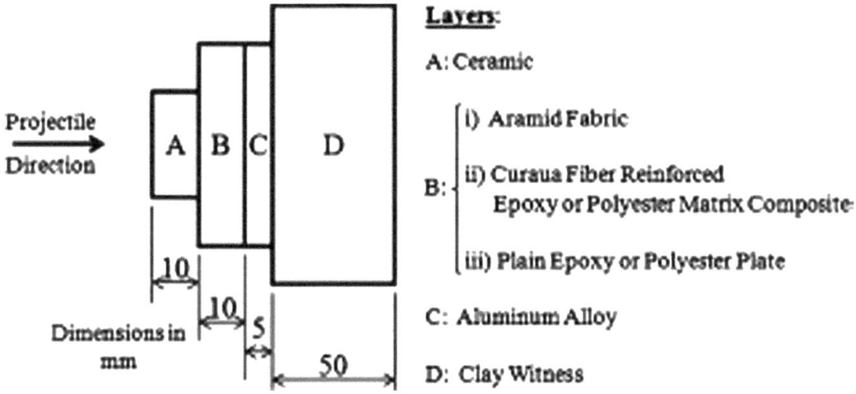

Fig. 2 Schematic diagram of the multilayered armor [7]

Based on the above considerations, the objective of the present work is to use arapaima scales as reinforcement phase in epoxy composites to be applied as part of a multilayered ballistic armor for protection against high velocity ammunition of rifle.

Materials and Methods

Figure 2 illustrates schematically the side view of the MAS arrangement used in this investigation [7].

The MAS investigated in the present work was composed of a front hexagon-Shaped (30 mm in edge and 10 mm in thickness) Al_2O_3–4%Nb_2O_5 ceramic tile. A rectangular plate (120 × 150 × 10 mm) was used as second layer of epoxy composites reinforced with arapaima scales. A back layer of 5052 H34 aluminum alloy, also as rectangular plate (120 × 150 × 5 mm), finished the MAS. The layers were bonded together using a polyurethane adhesive. The ceramic powders, Al_2O_3 (acquired from Treibacher Schleifmittel) and Nb_2O_5 (acquired from the Companhia Brasileira de Metalurgia e Mineração-CBMM) were mixed in a 4 wt% Nb_2O_5 proportion. The mixture was then Ball milled in aqueous suspension, dried at 60 °C for 48 h, sifted until the 0.355 screen, cold pressed at 30 MPa and sintered at 1400 °C for 3 h. Scales from the Brazilian company "Selva Amazônica" (body weight 100–150 kg). For the composite production, an epoxy resin hardened with 13 wt% triethylene tetramine (TETA), both supplied by the Epoxyfiber, Brazil, was used as polymeric matrix. The arapaima scales were planned and dried at 60 °C for 24 h for the production of composites. scales composite plates (120 × 150 × 10 mm) were prepared by compression molding, at room temperature (25 °C), in the volumetric fraction of 30 vol.%. Plain epoxy plates were similarly prepared. The scales were carefully positioned in the mold. After adding the resin with hardener, the mixture was kept under a pressure of

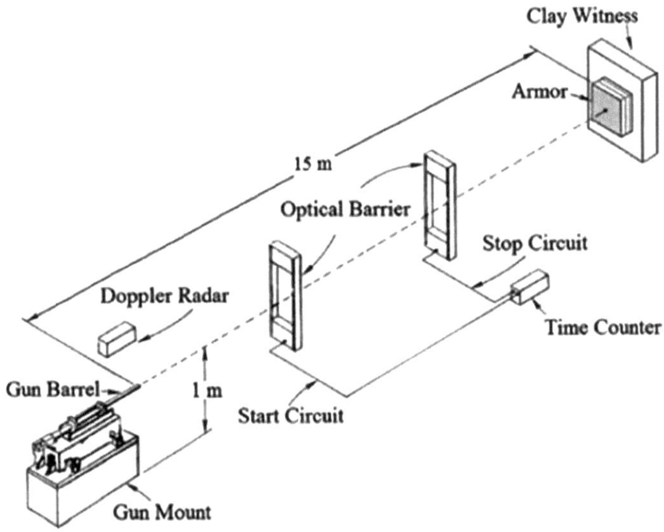

Fig. 3 Ballistic test: Schematic diagram with the experimental arrangement [8]

5 MPa for 24 h. Third MAS layer of the 5052 H34 aluminum alloy, with same 120 × 150 × 10 mm dimensions, were cut from sheets supplied by sheets were acquired from the Metalak Metais, Brazil. Ballistic tests were performed with NATO 7.62 × 51 mm military ammunition, following the NIJ 0101.06 procedures [8]. This means that the MAS target was positioned in front of a Roma Plastilina type of clay witness that simulates the consistency of the human body, 15 meters away from the shooting device. The experimental arrangement is shown in Fig. 3 and a real MAS mounted for the test in Fig. 4. The shooting device was a model B290 gun barrel with laser sight, produced by HPI - High Pressure Instrumentation, available at the Brazilian Army Assessment Center (CAEx), located in the Marambaia peninsula, Rio de Janeiro, Brazil. The bullet velocity was measured by both optical barriers and a model SL-520P Weibel doppler radar. The velocity was kept in the 847 ± 9.1 m/s range, measured 2.5 m before the target, as specified by the NIJ standard [8]. The indentation imprinted by the armor in the clay witness after the impact (backface signature) was measured by laser sensor and taken as a measure of the ballistic performance. The standard [8] specifies that the mean indentation be no deeper than 1.73 inch (44 mm).

Results and Discussion

Figure 4 shows the MAS target before and after the ballistic test. 10 indentation measurements were performed on the plastiline wall after impact. All indentation

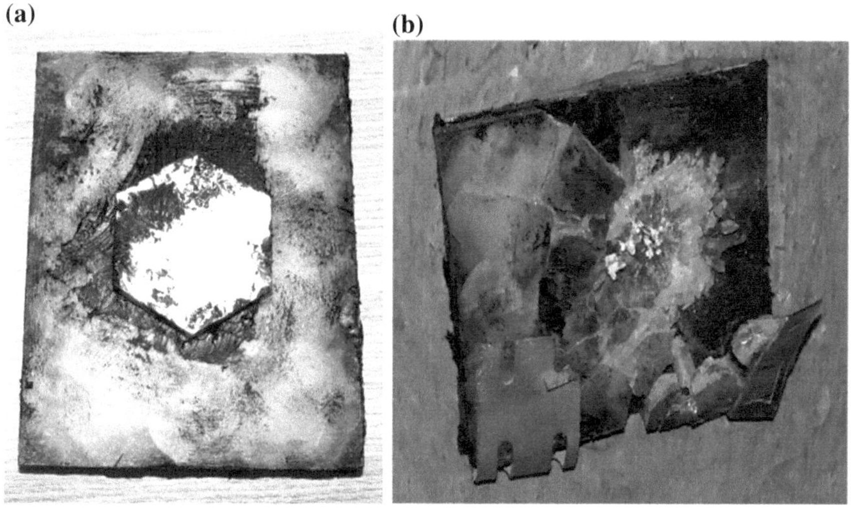

Fig. 4 The MAS before (**a**); and after the ballistic test Clay witness (**b**)

values were below 44 mm, and the tested MAS can be considered efficient according to the NIJ 0101.06 standard [8].

The data were statistically treated by the Weibull analysis, which gave information about the indentation distribution of probability. The Weibull's density of probability function is given by Eq. 1.

$$F(x) = 1 - exp\left[-\left(\frac{x}{\theta}\right)^{\beta}\right] \quad (1)$$

where x is the value assumed by the random variable; θ is the scale parameter and β is the shape parameter, also known as Weibull modulus. The Eq. 1 can be linearized for:

$$ln\left[ln\left(\frac{1}{1-F(x)}\right)\right] = \beta\ln(x) - \beta\ln(\theta) \quad (2)$$

The average indentation obtained was 14.14 ± 1.37 mm, which is well below the value of 44 mm required by the standard and even below Kevlar and other polymer composites studied, such as reinforced by curauá which is on average 23 mm. Using the Weibull analysis, the graph of Fig. 5 shows the probability density function of the indentation measurements.

To obtain the values of β and θ to linearize the graph through Eq. 2. The graph of Fig. 6 show this.

The high correlation coefficient (R^2 = 0,94) value indicate that does not change in the observed indentation ranges. In other words, the variability of depth of indentation values within the groups was purely statistical. the β parameter shows that the

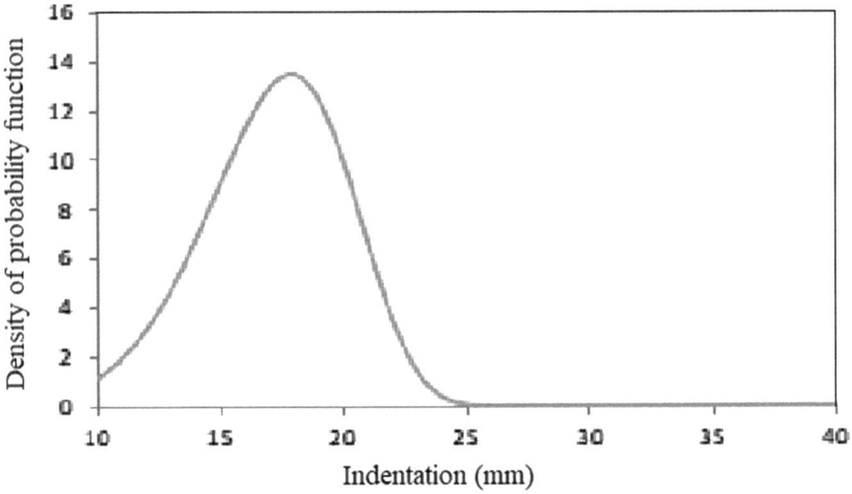

Fig. 5 Weibull density of probability function for the depths of indentation for the MAS

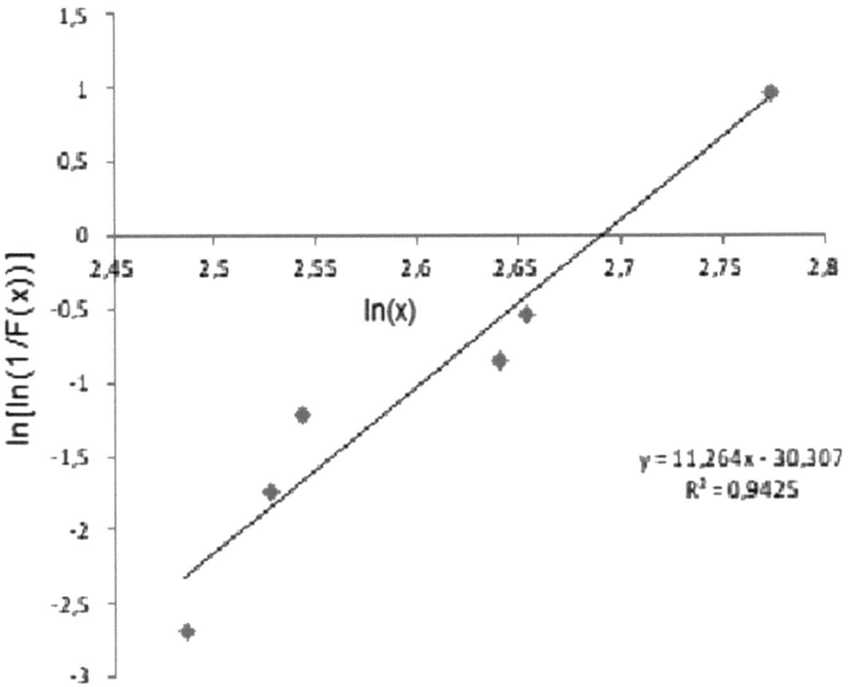

Fig. 6 Indents Weibull distribution graph

composite show very homogeneous results ($\beta = 11.26$). The θ parameter indicates the value of indentation for the MAS ($\theta = 14.74$) is better than that other works such as [9] where they obtained values of ($\theta = 23.00$) for MAS reinforced curaua fibers.

Conclusion

The composite reinforced with arapaima scales proved to be another green material that can potentially be used as ballistic armor that favour the environmental requirements.

Multilayered armor systems (MAS) using epoxy matrix composites reinforced with arapaima fish scales were considered to be equally efficient for the protection against 7.62 mm caliber ammunition, based on the NIJ 0101.06 criteria based on the depth of indentation in the clay witness (single-hit protection).

References

1. Drelich AJ, Monteiro SN, Brookins J, Drelish JW (2018) Fish scale: What can a surface innovator learn from this natural design? *Adv Mater*
2. Arola D, Murcia S, Stossel M, Pahuja R, Linley T, Devaraj A, Ramulu M, Ossa EA, Wang J (2018) The limiting layer of fish scales: Structure and properties. Acta Biomater 67:319–330
3. Lin YS, Wei CT, Olevsky EA, Meyers MA (2011) Mechanical properties and the laminate structure of Arapaima gigas scales. J Mech Behav Biomed Mater 4:1145–1156
4. Torres FG, Troncoso OP, Nakamatsu J, Grande CJ, Gómez CM (2008) Characterization of the nanocomposite laminate structure occurring in fish scales from Arapaima Gigas. Mater Sci Eng 28:1276–1283
5. Yang W, Sherman VR, Gludovatz B, Mackey M, Zimmermann EA, Chang EH, Schaible E, Qin Z, Buehler MJ, Ritchie RO (2014) Protective role of Arapaima gigas fish scales: Structure and mechanical behavior. Acta Biomater 10:3599–3614
6. Yang W, Chen IH, Gludovatz B, Zimmermann EA, Ritchie RO, Meyers MA (2013) Natural flexible dermal armor. Adv Mater 25:31–48
7. Monteiro SN, Louro LHL, Trindade W, Elias CN, Ferreira CL, Lima ES, Weber RP, Suarez JCM, Figueiredo ABS, Pinheiro WA, Silva LC, Júnior EPL (2015) Natural curaua fiber-reinforced composites in multilayered ballistic armor. Metall Mater Trans A, 8
8. National Institute of Justice–NIJ Standard 0101.04 (2000) Ballistic resistance of body armor. US Department of Justice–Office of Justice Programs, Washington, USA
9. Braga FO, Bolzan LT, Ramos FJHTV, Monteiro SN, Júnior EPL, Silva LC (2017) Ballistic efficiency of multilayered armor systems with sisal fiber polyester composites. Mater Res

Evaluation of Buriti Fabric as Reinforcement of Polymeric Matrix Composite for Ballistic Application as Multilayered Armor System

Luana Cristyne da Cruz Demosthenes, Lucio Fabio Cassiano Nascimento, Michelle Souza Oliveira, Fabio da Costa Garcia Filho, Artur Camposo Pereira, Fernanda Santos da Luz, Édio Pereira Lima, Jr., Leandro Alberto da Cruz Demosthenes and Sergio Neves Monteiro

Abstract Natural lignocellulosic fibers (NLFs) have been studied as a possible compound of multilayered armor system (MAS) when used as reinforcement in polymeric matrix composite. Among the NLFs, the buriti fiber ("*Mauritia Flexuosa*") is a highlight due to its mechanical properties such as tensile and flexural strength. Therefore, the use of these buriti fibers as fabrics in an epoxy composite is a promising candidate for ballistic against applications a high-velocity 7.62-mm ammunition as intermediate layers in the MAS. Composites reinforced with 10 vol% were evaluated and compared with aramid fabric as well as with other NLFs composites commonly used as second layer in the MAS. It was observed that the conditions studied exhibit a surprising behavior. Any of the samples were perforated after the ballistic impact. Moreover, the measured indentation depth is much lower than the maximum predicted by the international standard NIJ 0101.04.

Keywords Buriti fabric · Polymer composite · Ballistic performance · Multilayered armor system

Introduction

Humans throughout the ages used different types of materials in a way to promote protection against injuries from combat as well as dangerous situations. The armor concept can be classified depending on the intended applications such as body armor, light armor, or heavy armor [1].

L. C. da Cruz Demosthenes (✉) · L. F. C. Nascimento · M. S. Oliveira · F. da Costa Garcia Filho · A. C. Pereira · F. S. da Luz · É. P. Lima, Jr. · S. N. Monteiro
Military Institute of Engineering – IME, Rio de Janeiro, Brazil
e-mail: eng.luanademosthenes@gmail.com

L. A. da Cruz Demosthenes
Federal University of Amazonas – UFAM, Rio de Janeiro, Brazil

S. Ikhmayies et al. (eds.), *Green Materials Engineering*, The Minerals, Metals & Materials Series, https://doi.org/10.1007/978-3-030-10383-5_20

A recent increase of urban violence and armed conflict motivates researchers to create materials that withstand the most diverse artifacts and threats [1, 2]. Among these materials, the multilayered armor system (MAS) is considered as the most efficient ones against high-energy ammunition. The configuration of the MASs is, usually, composed of three distinct layers that have specific characteristics to resist the ballistic impact. The first layer is a hard and fragile material with the main objective to absorb most of the kinetic energy as well as to fragment the projectile [1–5]. As second layer materials with lower density than the ceramic, usually high-performance synthetic fibers such as the aramid fiber and the ultra-high-molecular-weight polyethylene are commonly employed. The third layer composed of an aluminum plate that has a ductile feature, capable of plastically deforming and retaining possible fragments, remnants of the projectile and previous layers [3].

In recent years, it has been investigated the possibility of the intermediary layer of the MAS be produced by polymeric composites reinforced with natural lignocellulosic fibers (NLF). These natural fibers have some advantages when compared to synthetic fibers such as lower specific weight and low toxicity as well as biodegradability, and low cost.

The buriti fibers ("*Mauritia Flexuosa*") are extracted from a palm tree typically of Amazonian and northeast forest floodplains. Buriti fibers' density is in the range from 0.63 to 1.12 g/cm^3 [6]. The chemical composition of this fiber is majorly composed by cellulose, about 51.3%, and lignin, about 16.4% [7]. As for the mechanical properties, tensile strength of 492.74 MPa for fibers with diameters ranging from 0.4 to 0.8 mm and modulus of elasticity of 24.87 GPa have been reported [6, 8]. Figure 1 presents the buriti palm tree (Fig. 1a), a bundle of fibers (Fig. 1b), and the buriti fiber fabric (Fig. 1c).

Materials and Methods

In order to produce the multilayered armor system in this research, ceramics based on aluminum oxide and epoxy matrix composites reinforced with 10 vol% of buriti fabric were fabricated. The ballistic ceramics were produced with alumina (96% w Al_2O_3), doped with niobium oxide (4% p Nb_2O_5) and sintered at 1,400 °C [9]. On the other hand, composite plates with dimensions of 120 × 150 × 10 mm^3 were fabricated with buriti fabric and epoxy resin-type diglycidyl ether of bisphenol A (DGEBA) hardener with triethylenetetramine (TETA) produced by Dow Chemical and supplied by the firm Epoxyfiber.

The MASs were finally prepared with the ceramic as the first layer, followed by the buriti epoxy-woven composite and, aluminum plate 5052 H34. Figure 2 shows the produced MAS.

The ballistic tests were conducted at the Brazilian Army Assessment Center (CAEx—Centro de Avaliação do Exército), Brazil. The ballistic test was carried out according to the NIJ 0101.04 standard using 7.62 × 51 mm M1 caliber (m = 9, 7 g). The MAS was fixed in clay witness was a Plastilina type, supplied by

Fig. 1 Buriti **a** palm tree, **b** detail of the possibilities of fiber extraction, and **c** buriti fabric used in the present work

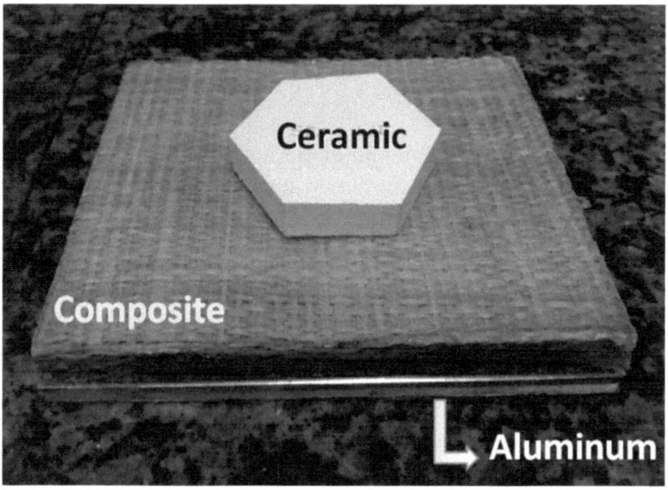

Fig. 2 MAS of buriti fabric

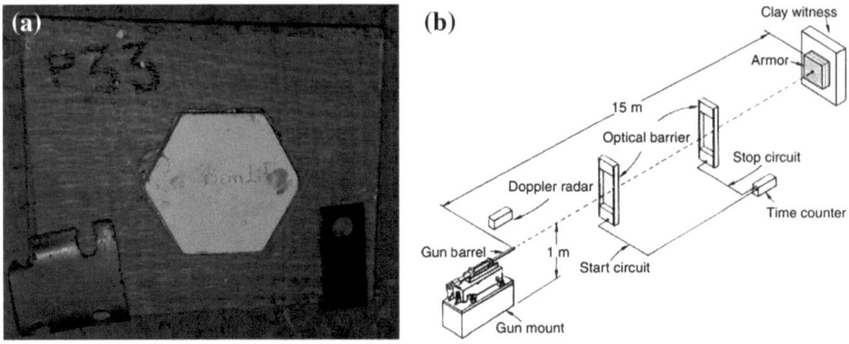

Fig. 3 MAS of buriti fabric essayed and **b** schematic exploded view of the ballistic experimental setup

Corfix (Brazil), applied to simulate the consistency of the human body. The shooting device was a model B290 high-pressure instrumentation (HPI) equipment, and the projectile's velocity measuring device was an optical barrier. According to the NIJ standard [10], the target is considered efficient as body armor if it is not perforated by the projectile and also do not present deep deformation on the clay witness. This deformation is called indentation and must be smaller than 1.73 in. (44 mm).

The surface of the fabric and the interaction zone between the projectile and the targets were observed using the Quanta FEG 250 FEG SEM in IME. The surfaces were sputtered coated to avoid charging (Fig. 3).

Results and Discussion

In all ballistic tests, 7.62-mm ammunition did not perforate the MAS target; i.e., it was always stopped before hitting the clay witness. Moreover, the indentation depth measured was two times lower than the maximum predicted by the NIJ standard (Table 1).

As expected, the ceramic first layer was fragmented in every test. Figure 4 presents the macroscopic aspect of the MAS before and after the ballistic impact.

Figure 5 shows the cloud of ceramic fragments was captured by fabrics after ballistic test. Figure 5a represents aramid fabric [11], and Fig. 5b represents buriti fabric both after ballistic impact. It is possible to observe that the two materials have a characteristic of absorbing the ceramic fragment.

In order to implement a qualitative analysis of these samples, the analysis of dispersive energy spectroscopy (EDS) was carried out. The objective was to verify the possible chemical elements present in the highlighted region of Fig. 6b. The data obtained were applied in Table 2.

Table 1 Indentation for the different MAS

Second layer material	Indentation	References
Epoxy reinforced with 10% buriti fabric	22 ± 4	PW[a]
Epoxy reinforced with 30% curaua non-woven fiber fabric	28 ± 3	[11]
Epoxy reinforced with 30 vol% aligned curaua fiber	22 ± 3	[12]
Aramid fiber fabric laminate	21 ± 3	[13]
Epoxy composite reinforced with 30% of jute fabric	21 ± 3	[3]

[a]Present work

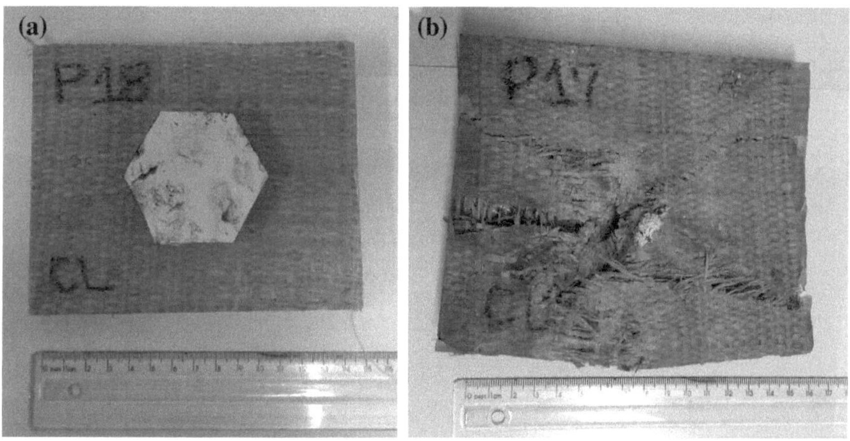

Fig. 4 MAS buriti fabric **a** before and **b** after ballistic test

Fig. 5 Micrograph of samples MAS after ballistic test: **a** Aramida fabric (5000×) [11] and **b** buriti fabric (1300×)

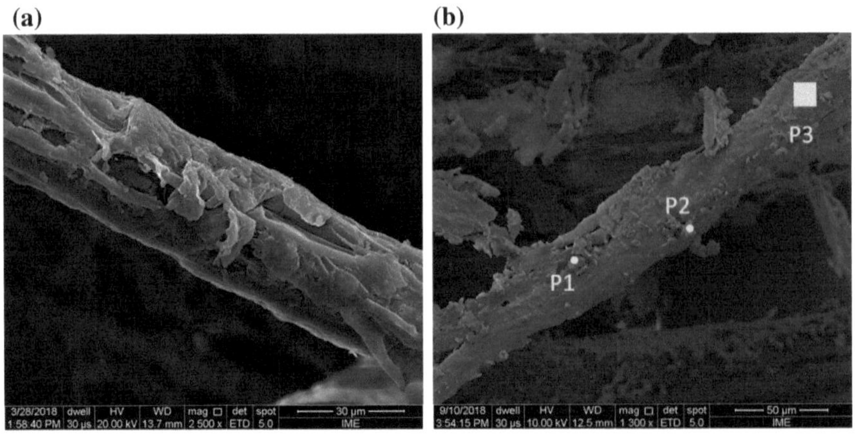

Fig. 6 SEM of buriti fabric **a** before impact (2500×) and **b** after impacted with highlight points the analysis chemical composition (1300×)

Table 2 Chemical composition in highlighted points of the Fig. 6b

Elements	Composition (%)		
	Point 1	Point 2	Point 3
Au	49.08	15.62	25.79
C	34.00	51.89	36.06
O	13.57	11.07	10.30
Pb	2.34	0	0
Al	1.01	12.91	14.66
Nb	0	8.51	11.22

The gold element, present in Table 2, refers to the coating used in the samples. The samples were covered to make the conductive surface and finally improve the signal in the SEM. Elements as carbon and oxygen are characteristic of natural fibers. Others elements such as alumina and niobium oxide were elements correspond to the ceramic particles that fixed in that region. Furthermore, it is possible to observe the presence of lead, which is a remnant of the projectile impregnated in the region.

Summary and Conclusions

An epoxy matrix composite reinforced with buriti fabric in substitution for conventional aramid fabric, with same total thickness, in MAS for personal protections attended the NIJ trauma limit after ballistic tests with a high-velocity 7.62 × 51-mm ammunition. In this way, there were not any perforations on the ballistic test, and the indentation was 1.73 in. (44 mm) lower.

Reinforced composites with 10% buriti fabric presented indentations similar to those of other natural fiber fabrics with 30% reinforcement. Therefore, it is believed that the use of the composite with values of 30% is more favorable.

The micrographics of buriti fabrics have the same characteristic of the aramid in terms of retaining the ceramic particles after impact. Despite that, the composite was not able to keep its integrity after the ballistic impact and this is one of the main parameters to measure ballistic efficiency.

Acknowledgements The authors of the present work thank the Brazilian supporting agencies CAPES, CNPq, and FAPERJ for the funding and the CAEx for performing the ballistic tests.

References

1. Braga FO, Bolzan LT, Lima EP, Monteiro SN (2017) Performance of natural curauá fiber-reinforced polyester composites under 7.62 mm bullet impact as a stand-alone ballistic armor. J Mater Res Technol 6(4):323–328
2. Medvedovski E (2010) Ballistic performance of armour ceramics: influence of design and structure - Part 1. CeramicsInternational 36:2103–2115
3. Luz FS, Lima EP, Louro LHL, Monteiro SN (2015) Ballistic test of multilayered armor with intermediate epoxy composite reinforced with jute fabric. Mater Res 18:170–177
4. Reportagem Jornal Nacional, Rio tem 30 PMs mortos em 3 meses; cabo morre baleado na Pavuna. G1 2018. [acesso em 7 jun. 2018] 22:45. Disponível em: http://g1.globo.com/jornal-nacional/noticia/2018/03/rio-tem-30-pms-mortos-em-3-meses-cabo-morre-baleado-na-pavuna.html
5. Nascimento LFC, Holanda LIF, Louro LHL, Monteiro SN, Gomes AV, Lima EP (2017) Natural mallow fiber-reinforced epoxy composite for ballistic armor against class III-A ammunition. Metall Mater Trans A 48(10):4425–4431
6. Monteiro SN, Lopes FPD, Barbosa AP, Bevitori AB, Silva ILA, Costa LL (2011) Natural lignocellulosic fibers as engineering materials—an overview. Metall Mater Trans A 42(10):2963–2974
7. Barbosa AP (2011) Características estruturais e propriedades de compósitos poliméricos reforçados com fibras de Buriti. Tese (Doutorado em Engenharia e Ciência dos Materiais) - Universidade Estadual do Norte Fluminense Darcy Ribeiro
8. Santos NSS (2010) Análise experimental e teórica do comportamento mecânico sob carregamentos quase-estáticos de compósitos reforçados com fibras vegetais. Tese (Doutorado em Engenharia Mecânica) - Universidade Estadual de Campinas, Campinas, SP
9. Santos JL, Marçal RLSB¸ Jesus PRR, Gomes AV, Lima EP, Monteiro SN, De Campos JB, Louro LHL (2017) Effect of LiF as sintering agent on the densification and phase formation in Al_2O_3-4 Wt Pct Nb_2O_5 ceramic compound. Metall Mater Trans A 48:4432–4440
10. NIJ Standard 0101.04 (2000) Ballistic resistance of personal body armor. US Department of Justice/Office of Justice Programs, National Institute of Justice
11. Braga FO, Bolzan LT, Luz FS, Lopes PHLM, Lima EP Jr, Monteiro SN (2017) High energy ballistic and fracture comparison between multilayered armor systems using non-woven curaua fabric composites and aramid laminates. J Mater Res Tecnol 6(4):417–422
12. Monteiro SN, Louro LHL, Trindade W, Elias CN, Ferreira LF, de Lima ES, Weber RPW, Suarez JCM, da Figueiredo ABHS, Pinheiro WA, da Silva LCS, Lima EP (2015) Natural curaua fiber-reinforced composites in multilayered ballistic armor. Metall Mater Transcr A 46(10):4567–4577
13. Monteiro SN, Milanezi TL, Louro LHL, Lima EP, Braga FO, Gomes AV et al (2016) Novel ballistic ramie fabric composite competing with kevlar^{TM} fabric in multilayered armor. Mater Des 96:263–269

Evaluation of the Absorbed Energy and Velocity Limits of Reinforced Epoxy Composites with Mallow Natural Fibers Used in Ballistic Protection

Lucio Fabio Cassiano Nascimento, Sérgio Neves Monteiro, Jheison Lopes dos Santos, Ulisses Oliveira Costa and Luana Cristyne da Cruz Demosthenes

Abstract In the present work, natural fibers of mallow (*Urena Lobata, Linn*) were used in percentages of 0, 10, 20, and 30% vol. for ballistic application in epoxy matrix composites. The ballistic efficiency of these composites was evaluated through the measurement of the absorbed energy and the velocity limit, after impact of 7.62 mm ammunition, in order to compare with work that used composites with other natural fibers and traditional materials, such as aramid fabrics, used in vests for individual protection. The results showed through visual analysis and scanning electron microscopy (SEM) indicate the main mechanism of rupture acting in the composites was the delamination of layers. In all cases analyzed in this work, the parameters obtained for the residual velocity test were higher than those found for aramid tissue. This fact evidences the viability of the fibers/mallow fabric for use in dynamic applications, especially those related to ballistic protection.

Keywords Ballistic test · Mallow fibers · Composites

Introduction

Tripod formed by mobility, resistance to penetration and high absorption of impact are the main factors that make ballistic armoring efficient. However, improvement in one of these factors often negatively influences that of another. For example, an increase in the penetration resistance may cause a reduction in mobility, due to the necessary increase of thickness and, therefore, of weight, in the case of monolithic armors made of steel [1]. There have been general needs to seek new engineering solutions to improve the protection of the targets. Armors constituted by ceramic components associated with other materials, such as polymer composites, are then employed to favor a good ballistic protection/weight ratio [2]. These are called multilayer armoring systems (MAS).

L. F. C. Nascimento (✉) · S. N. Monteiro · J. L. dos Santos · U. O. Costa · L. C. da Cruz Demosthenes
Military Institute of Engineering—IME, Rio de Janeiro, Brazil
e-mail: lucio_coppe@yahoo.com.br

S. Ikhmayies et al. (eds.), *Green Materials Engineering*, The Minerals, Metals & Materials Series, https://doi.org/10.1007/978-3-030-10383-5_21

Table 1 Properties of aramid fiber, mallow fiber, and epoxy resin [2, 7–10]

Materials	Density (g/cm^3)	Tensile resistance (MPa)	Young's modules (GPa)	Specific resistance (MPa)	Specific modules (GPa)
Aramid (Kevlar™)	1.4	3000–3150	63–67	2143–2250	45–48
Mallow	1.4	160	17.4	116	13
Epoxy	1.1–1.3	60–80	2–4	46–73	1.5–3.6

However, synthetic materials, such as Kevlar™, that make up the composite of the intermediate layer have drawbacks, such as high acquisition and processing costs, as well as environmental damages, due, for example, to their degradation over time, after the end of its useful life. As a result, constant research efforts are looking for new alternative materials that can efficiently replace synthetic ones and are environmentally friendly. In this context, natural lignocellulosic fibers (NLFs) can play this role because they have many advantages, such as: good specific properties (low specific mass), low toxicity, biodegradability, and low cost. In addition, they make it possible to foster the economic and technological development of Brazil's less favored regions, such as the northern and northeastern regions, which are major producers of natural fibers. Other properties can be found in Table 1.

The goal of this study is to analyze the ballistic efficiency of epoxy matrix composites in different percentage of mallow fibers (0, 10, 20, and 30% vol.). The ballistic test consists of determining the residual velocity of the bullet after its impact on the target (specimen). This velocity is related to the energy that is absorbed by the composite during the impact—the greater that energy, the more efficient the composite will be. After, a visual analysis is made in scanning electron microscopy (SEM) to observe fracture mechanisms in the composites.

Materials and Methods

The mallow fiber has been used as reinforcement in the composite. The fibers were provided by State University of Northern Rio de Janeiro (UENF). Firstly, the fibers were cleaned, aligned, and cut in length of 15 cm. After, they were maintained in stove for 24 h in order to reduce the absorbed moisture. The as received bundle of fibers and the mallow and jute hybrid fabric are shown in Fig. 1.

The composite matrix was diglycidyl ether of the bisphenol A (DGEBA), which is a commercial epoxy resin mixed with triethylenetetramine (TETA) hardener in proportion of 13 parts of hardener per hundred parts of resin in weight. Both components are fabricated by the Brazilian firm Dow Chemical and commercially supplied by the distributor RESINPOXY Ltda.

Composites with 0, 10, 20, and 30% in volume of mallow fibers were manufactured (Fig. 2), being eight units of each concentration. It was used to mallow fibers as initial

Fig. 1 Mallow fibers as received

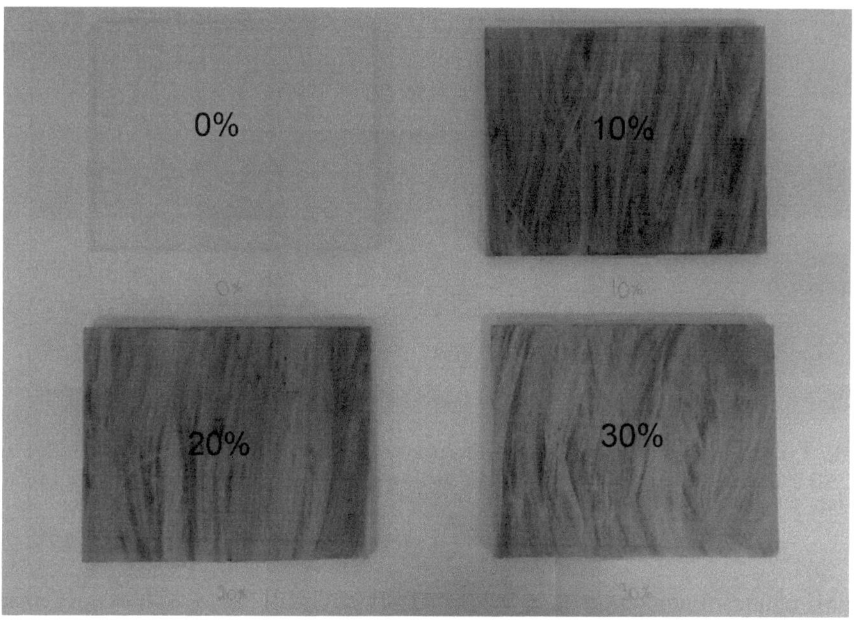

Fig. 2 Composite plates of epoxy reinforced with 0, 10, 20, and 30% vol. of mallow fibers

reference density of 1.40 g/cm^3 [3] and for the epoxy resin (DGEBA-TETA) value of 1.11 g/cm^3 [4]. A pressure of 5 MPa was applied and the composite plate cured for 24 h [2, 4].

The ballistic tests were conducted at the Brazilian Army shooting range facility, CAEX, in the Marambaia peninsula, Rio de Janeiro. All tests, eight for each type of percentage, were carried out according to the NIJ 0101.04 standards using 7.62 mm

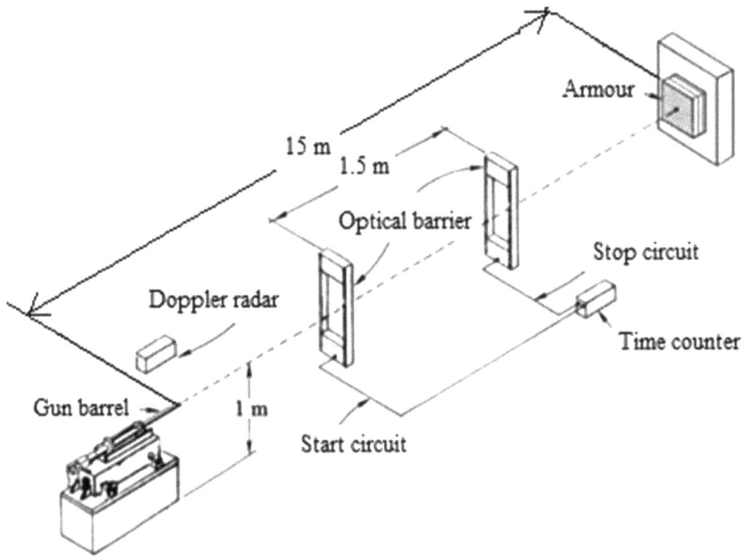

Fig. 3 Schematic exploded view of the ballistic experimental setup [2]

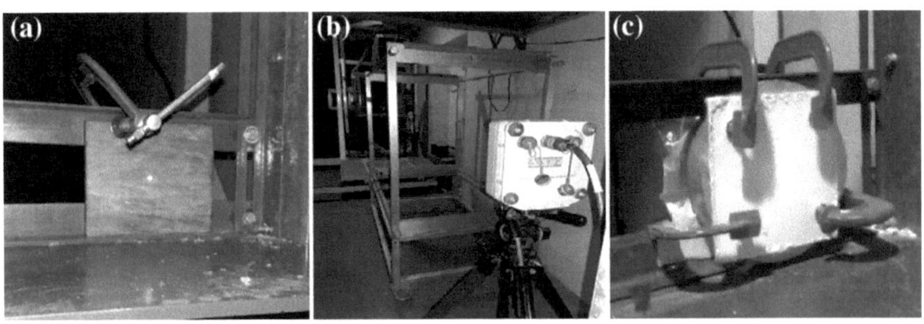

Fig. 4 **a** Specimen of composite plate fixed in the metal bracket with spring clips. **b** Apparatus of CAEX for ballistic testing with ammunition and doppler radar. **c** Aramid plate fixed with spring clips

FMJ military ammunition (m = 9.7 g—Armor level III). Figure 3 shows, schematically, the exploded view of the ballistic test setup. A dashed straight line indicates the projectile trajectory. A steel frame was used to position the target, which was held in place by spring clips (Fig. 4a). The gun, located 15 m from the target, was sighted on its center with a laser beam. The exact velocity of the projectile at three moments: leaving the gun, immediately before impacting (V_i) in the plate and after outcoming projectile (residual velocity—V_r) was measured by an optical barrier and a model SL-52 OP Weibel fixed-head Doppler radar system (Fig. 4b). The Kinetic energy ΔE_d, dissipated inside the target, could then be estimated by the Eq (1):

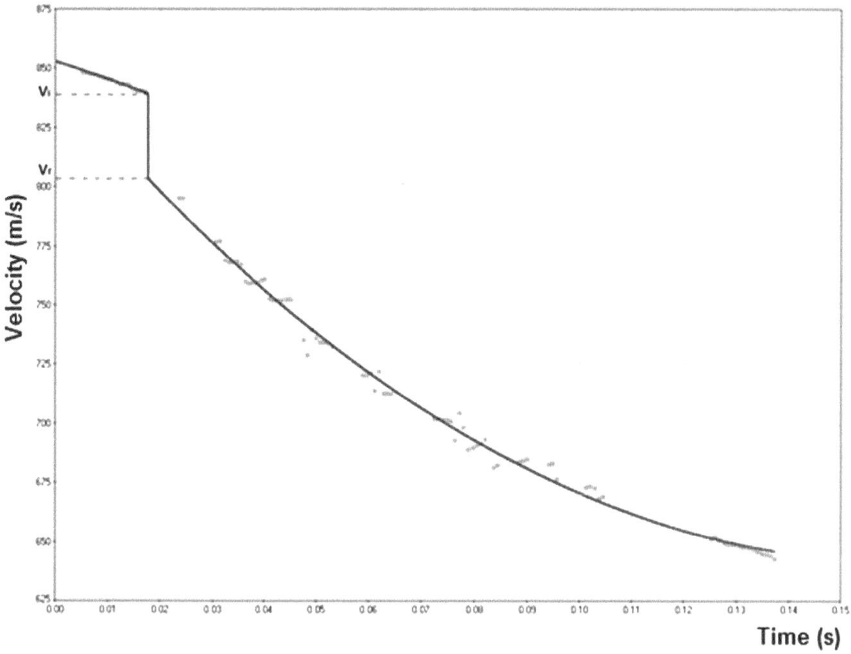

Fig. 5 Graph of experimental points of the reinforced epoxy composite with 30% vol. of mallow fibers

$$\Delta E_d = \frac{1}{2}m(V_i^2 - V_r^2) \quad (1)$$

Results and Discussion

In order to verify the individual ballistic resistance of each component of the MAS, the residual velocity test was performed, and it was possible to estimate the absorbed energy (E_{abs}) and the velocity limit (V_L) of each sample with the aid of a radar (Fig. 4b) and using Eq. (1). In this test, all the samples were perforated and crossed by the projectile, which allowed the measurement of the energy absorbed by each component present in the shield. Figure 5 shows an example of data obtained from the Doppler radar spectrum of a test specimen of the epoxy composite mallow fibers (30% vol.) and the adjusted continuous polynomial curve.

Approximately 840 m/s, there was an abrupt drop in velocity, which characterized the moment of impact on the target. This velocity was defined as the velocity of impact (V_i) and the minimum velocity reached at that fall defined the residual velocity (V_r). Similar graphs were obtained for the other components of the MAS and based on the results extracted from these graphs both the limit velocity (V_L)

Table 2 Impact and residual velocities together with internally dissipated energy in ballistic tested to composites reinforced with mallow fibers [2, 4–6]

Armor components	V_i (m/s)	V_r (m/s)	E_{abs} (J)	% E_{abs}	V_L (m/s)	Reference
Aramid fabric (16 layers)	862 ± 7.00	835 ± 10.00	221.00	6.10	212 ± 23	[6]
Composite 100% epoxy	848.08 ± 6.20	817.01 ± 8.87	250.73	7.20	227.02 ± 12.52	Present work
Composite 100% mallow (30% v/v)—Fibers	840.00 ± 11.19	807.96 ± 11.52	256.02	7.48	229.54 ± 10.06	Present work
Composite 100% mallow (20% v/v)—Fibers	843.39 ± 5.33	807.07 ± 8.57	290.53	8.42	243.81 ± 21.42	Present work
Composite 100% mallow (10% v/v)—Fibers	845.02 ± 6.78	815.09 ± 9.78	240.72	6.95	220.92 ± 28.77	Present work

and the absorbed energy (E_{abs}) were determined. The results of the average impact velocity of the projectile (V_i), mean residual velocity (V_r), absorbed energy (E_{abs}) and their percentage in relation to total energy (% E_{abs}) are shown in Table 2, as well as the estimated velocity limit (V_L). For comparative purposes, data from other authors who also studied MAS based on natural fibers were also presented in Table 2 [2, 4–6].

The values of the boundary velocity (V_L) and absorbed energy (E_{abs}) for epoxy matrix composites reinforced with 20% v/v of mallow fibers presented values higher than those found for aramid [6] and other percentages of reinforcement. This fact can be explained by the good mechanical properties of mallow, as shown in Table 1. There was dissipation of a greater amount of kinetic energy of the projectile through the fragile fracture mechanisms of the epoxy matrix, together with ductile mechanisms such as delamination between layers, elastic deformation of the composite, shear of the layers, and stress at fiber rupture [2]. In all cases analyzed by this work, the parameters obtained for the residual velocity test were higher than those found for aramid fabric (212 m/s and 221 J), which shows the viability of mallow fibers for use in ballistic applications.

Figure 6a, b show that the mechanisms associated with fragile fracture are more active in the system, as a function of the percentages of 0 and 10% v/v reinforcement. The main working mechanisms were those corresponding to fragile and catastrophic fracture, typified by the cleavage of planes and river marks. In Fig. 6c,d the ductile fracture mechanisms control the fracture, as a function of the percentages of 20 and 30% v/v reinforcement. SEM images clearly demonstrated the similarity in the main fracture mechanisms acting in the process: layer delamination, fiber fracture and pull-out. Probably the energy absorbed with 20% v/v was superior to all other

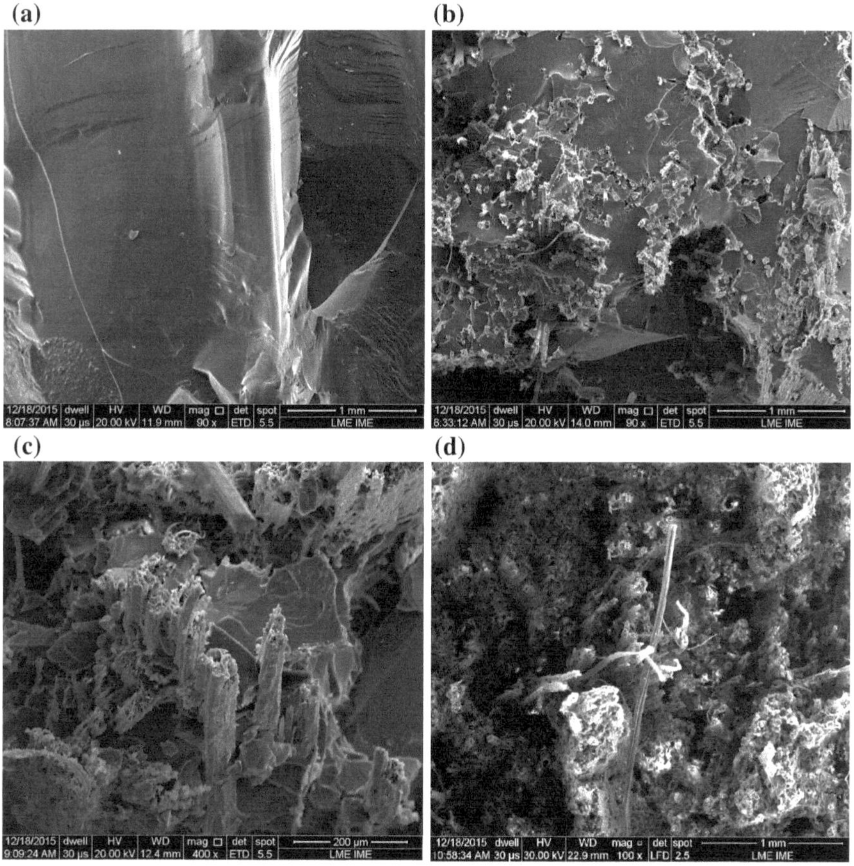

Fig. 6 **a** Composite 100% epoxy (90×); **b** Composite epoxy reinforced with 10% v/v of mallow fibers (90×); **c** Composite epoxy reinforced with 20% v/v of mallow fibers (400×); **d** Composite epoxy reinforced with 30% v/v of mallow fibers (100×)

reinforcement percentages, due to associated with the ideal wettability of the fibers by the epoxy matrix.

Conclusions

SEM shows that the mallow fiber contributed in an effective way to increase the resistance of the epoxy matrix through the mechanisms of energy absorption, such as delamination of layers, detachment, and disruption of fibers.

The values of the velocity limit (V_L) and absorbed energy (E_{abs}), for epoxy matrix composites reinforced with 20% v/v of mallow fibers presented higher values than those found for aramid [6] and other percentages of reinforcement. In all cases analyzed by this work, the parameters obtained for the residual velocity test were higher than those found for the aramid fabric (212 m/s and 221 J), being that these values grow with the increase of the percentage of fiber reinforcement used, which shows the viability of the mallow fibers for use in ballistic applications.

Acknowledgements The authors acknowledge the support to this investigation by the Brazilian agencies CNPq, CAPES, FAPERJ, and UENF for supplying the mallow fibers.

References

1. Borvik T, Langseth M, Hopperstad OS, Malo KA (1999) Ballistic penetration of steel plates. Int J Impact Eng 22:885–886
2. LUZ FS (2014) Ballistic behavior evaluation multilayered armour with epoxy composite reinforced with jute fiber (dissertation). Military Engineering Institute, Rio de Janeiro. (in Portuguese)
3. Oliveira JTS (1998) Characterization of eucalyptus wood for construction (thesis). Polytechnic School of the University of São Paulo, São Paulo. (in Portuguese)
4. da Silva LC (2014) Ballistic behavior of natural fiber epoxy composite multi-layer armour (thesis). Military Engineering Institute, Rio de Janeiro. (in Portuguese)
5. de Araújo BM (2015) Evaluation of ballistic behavior of multi-layered armouring with epoxy composite reinforced with sisal fiber (dissertation). Military Engineering Institute, Rio de Janeiro. (in Portuguese)
6. Braga FO (2015) Ballistic behavior of a multi-layered armouring using polyester-curauá composite as intermediate layer (dissertation). Military Engineering Institute, Rio de Janeiro. (in Portuguese)
7. Medvedovski E (2010) Ballistic performance of armour ceramics: influence of design and structure-part 1. Ceram Int 36:2103–2115
8. Lee YS, Wetzel ED, Wagner NJ (2003) The ballistic impact characteristic of kevlar woven fabrics impregnated with a colloidal shear thickening fluid. J Mater Sci 38:2825–2833
9. Nascimento LFC (2017) Characterization of the epoxy-mallow fiber composite for use in multilayer ballistic armouring (thesis). Military Engineering Institute, Rio de Janeiro. (in Portuguese)
10. Margem JI (2013) Analysis of structural characteristics and properties of polymer composites reinforced with mallow fibers (thesis). State University of Norte Fluminense, Rio de Janeiro. (in Portuguese)

Fique Fiber-Reinforced Epoxy Composite for Ballistic Armor Against 7.62 mm Ammunition

Michelle Souza Oliveira, Artur Camposo Pereira, Fabio da Costa Garcia Filho, Fernanda Santos da Luz, Fabio de Oliveira Braga, Lucio Fabio Cassiano Nascimento, Édio Pereira Lima, Jr., Luana Cristyne da Cruz Demosthenes and Sergio Neves Monteiro

Abstract Multilayered armor system (MAS) layers are composed of materials such as a front ceramic and a back composite that must show both high impact resistance and low weight. Recently, composites reinforced with natural fibers have been considered as MAS second layer thanks to their efficient ballistic performance associated with other advantages such as being cheaper and environmentally friendly. Among the natural fibers, the fique fiber has shown potential as reinforcement of polymer composites for engineering applications. Epoxy matrix composites reinforced with up to 25 vol.% of fique fiber were for the first time ballistic tested. For practical application as armor for personal protection, the layer of fique fiber composite presents not only a superior ballistic performance but also lightness and economical advantages over the conventional aramid fabric.

Keywords Fique fiber · Epoxy matrix · Ballistic test · Multilayered armor

Introduction

Violence and armed conflict require effective systems of personal protection. For high caliber ammunition, such as 7.62 mm, a single material may not be sufficient to stop the projectile and prevent a lethal trauma to the armor user. For this type of weaponry, lightweight and efficient solutions have been researched and developed, such as the multilayered armor systems (MASs) [1–5].

M. S. Oliveira (✉) · A. C. Pereira · F. da Costa Garcia Filho · F. S. da Luz · L. F. C. Nascimento · É. P. Lima, Jr. · L. C. da Cruz Demosthenes · S. N. Monteiro
Military Institute of Engineering - IME, Rio de Janeiro, Brazil
e-mail: oliveirasmichelle@gmail.com

F. de Oliveira Braga
National Service of Industrial Apprenticeship - SENAI, Rio de Janeiro, Brazil

F. de Oliveira Braga
Fluminense Federal University - UFF, Rio de Janeiro, Brazil

S. Ikhmayies et al. (eds.), *Green Materials Engineering*, The Minerals, Metals & Materials Series, https://doi.org/10.1007/978-3-030-10383-5_22

These systems can be used for ballistic protection in various applications, such as in personal vests, vehicles, aircraft, and large structures. The front layer of the MAS is commonly made of ceramic material. This ceramic material has the ability to erode and deform the front of projectile, and its fragmentation is an efficient mechanism for dissipating the kinetic energy of the projectile. As second layer, composites or laminated fabrics such as Kevlar® are used, whose main function is to absorb another portion of energy by capturing the fragments generated from the ceramic fragmentation as well as from the projectile [6–17]. A third layer of light and ductile metal may be considered. The second layer deserves special attention. The lower its density, the lower the compression wave intensity that reaches the user of the armor, providing greater security. The wave portion that is reflected at the interface between the materials, in this case, causes the ceramic to fragment, which increases the energy dissipation [12]. Polymeric composites reinforced with natural fibers are good alternatives, since they have good resistance, low density, and consequent good performance [18]. In addition, they are low cost and environmentally friendly. Several papers have been published with the purpose of investigating the behavior of these composites as components of MAS [3, 4, 6–16]. Among the less-known lignocellulosic fibers, the fique fiber has a potential for composites reinforcements, and the application of its composites in MAS is promising. To the date, no work has been devoted to the study of the behavior of epoxy matrix composites reinforced with fique in MAS. Therefore, in this work, it was investigated the behavior of reinforced epoxy composites with 5, 15, and 25% fique in MAS with ceramic front layer and aluminum alloy backsheet.

Materials and Methods

Figure 1 illustrates schematically the 3D view of the multilayered armor system (MAS) arrangement used in this investigation.

An ultraflex-brand polyurethane adhesive joined the layers. Al_2O_3 powders (supplied by Treibacher Schleifmittel, Brazil) and Nb_2O_5 (supplied by CBMM, Brazil) were mixed in fixed weight proportion of 96:4 with LiF (0.50 wt%) and polyethylene glycol (PEG), and as binder, both were acquired from Vetec, Brazil. The mixture was milled, sieved, cold-pressed, and sintered at 1400 °C for 3 h. The polymeric matrix was epoxy diglycidyl ether of bisphenol-A (DGEBA) resin, hardened with triethylene tetramine (TETA, 13 phr), both produced by Dow Chemical and acquired from Brazilian company Epoxyfiber, and also the fiber (supplied by Compañia de Empaques, Itagüaí, Antioquia, Colombia) previously cleaned and dried (2 h, 60 °C) in a steel matrix. The mixture was maintained under pressure of 3 MPa until curing (25 °C, 24 h). The 5-mm thick aluminum plate was supplied by Metalak, Brazil.

Ballistic tests were carried out at the Brazilian Army shooting range facility, CAEX, in the Marambaia peninsula, Rio de Janeiro.

The samples were positioned in front of a block of clay witness, which simulates the consistency of the human body. The clay witness was a commercially supplied

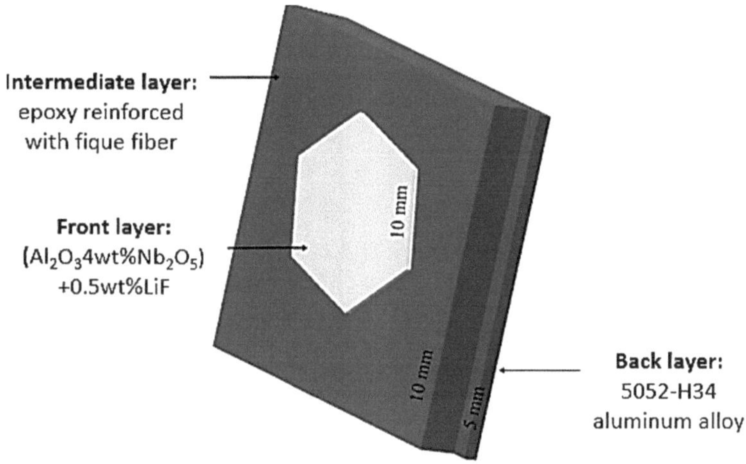

Fig. 1 Multilayered armor target schematic assembly

CORFIX® plastilene. After the ballistic impact, the depth of the deformation caused in the block was measured, which according to NIJ 0101.04 [19] should have a maximum of 44 mm for which the MAS can be considered efficient.

Results and Discussion

The depth of penetration (DOP) in ballistic test, corresponding to the portion of undissolved energy related to the trauma caused by the mass block, which was measured by means of a laser sensor and is shown in Table 1.

It is observed that the MAS was efficient with respect to the criterion of NIJ 0101.04 [19], both exhibited an indentation (DOP) less than 44 mm. Considering the standard deviations, the values presented in Table 1 were very close (especially for 5 and 15 wt%), indicating that the different MASs have similar ballistic efficiency.

Figure 2 shows the armor arrangements before and after the ballistic impact.

Table 1 Average depth of penetration in the clay witness backing MAS of fique fabric composite

Intermediate layer material	Depth of penetration (mm)
Epoxy composite reinforced with 5 vol.% fique fiber	21.8 ± 0.13
Epoxy composite reinforced with 15 vol.% fique fiber	21.8 ± 0.66
Epoxy composite reinforced with 25 vol.% fique fiber	18.9 ± 0.25
Plain epoxy plate	20.0 ± 1.00 [5]
Kevlar	23.0 ± 3.00 [5]

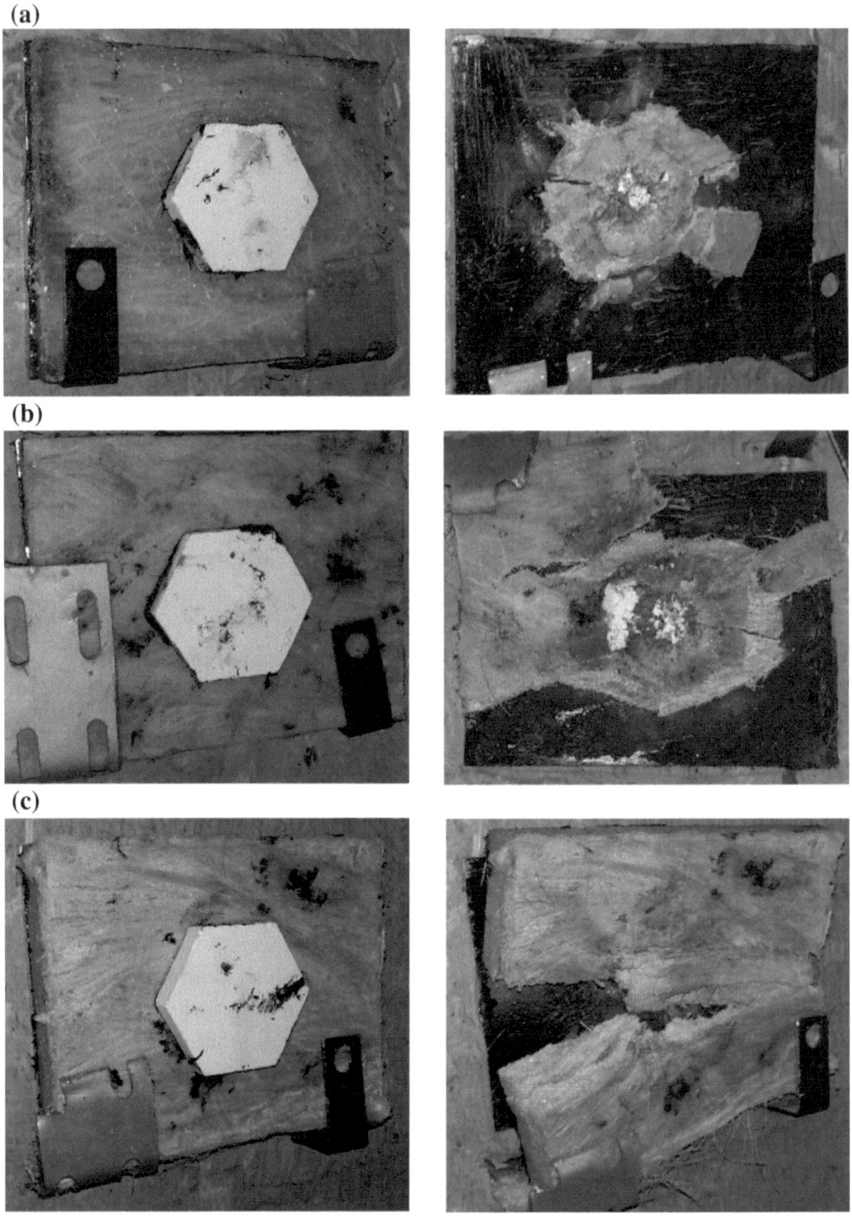

Fig. 2 MAS targets of **a** 5 wt%, **b** 15 wt%, and **c** 25 wt% composite after ballistic test

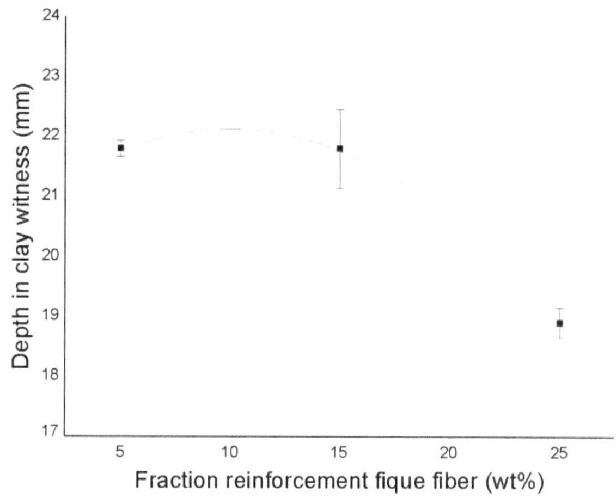

Fig. 3 Depth of penetration (DOP) after impact against the targets as a function of the fraction reinforcement fique fiber

Delamination seems to be one of the main mechanisms of fracture, as it was observed in most of the tests, independently of the fiber content. It consists of the separation of composite phases, which together with other mechanisms such as fragile fracture of the polymer matrix, rupture of the fibers, and separation of fibers in microfibrils play a fundamental role in dissipation of the kinetic energy of projectile.

It was observed that all the composites presented satisfactory depth of deformation, but both lost their integrity after the impact.

Figure 3 shows that composite with higher fraction reinforcement fique fiber (25 wt%) performs better as the indentation recorded. Additionally, Monteiro et al. [9] presented the economic advantage about use of armor vests with fique fiber composites, which would be 13 times cheaper than similar ones made with Kevlar™.

Summary and Conclusions

- Epoxy matrix composite reinforced with fique fiber used in substitution for conventional aramid fabric plates, with same total thickness, in a multilayered armor for personal protections attended the NIJ trauma limit after ballistic tests with high velocity 7.62 × 51 mm ammunition.
- The different MAS showed similar ballistic efficiency; this may be related to the ability of the different composites to dissipate the impact by means of the main mechanism related to both the polymer matrix and the fibers, known as delamination.
- The partial disintegration of the fique fiber composites after the MAS ballistic impact limits their application to small mosaic pieces, especially 5 and 15 wt%, to stand against multiples shots of 7.62 mm projectiles.

- In principle, both technical and economical reasons support the replacement of aramid fabric, as second layer backing the front ceramic in a mass, by fique reinforced epoxy composite, in which the natural fiber is also environmentally friendly.

Acknowledgements The authors thank the Brazilian agencies CAPES and CNPq for the financial support and CAEX for performing the ballistic tests.

References

1. MedvedovskI E (2010) Ballistic performance of armour ceramics: influence of design and structure. Part 1. Ceram Int 36:2103–2115
2. Akella K, Naik NK (2015) Composite armour—a review. J Indian Inst Sci 95(3):297–312
3. Braga F, Bolzan LT, Luz FS, Lopes PH, Lima EP Jr, Monteiro SN (2017) High energy ballistic and fracture comparison between multilayered armor systems using non-woven curaua fabric composites and aramid laminates. J Mater Res Technol 6(4):417–422
4. Cruz RB, Lima EP Jr, Monteiro SN, Louro LH (2015) Giant bamboo fiber reinforced epoxy composite in multilayered ballistic armor. Mater Res 18(Suppl 2):70–75
5. Loverro KL, Brown TN, Coyne ME, Schiffman JM (2015) Use of body armor protection with fighting load impacts soldier performance and kinematics. Appl Ergon 46:168–175. https://doi.org/10.1016/j.apergo.2014.07.015
6. Luz FS, Lima Jr. EP, Louro LH, Monteiro SN (2015) Ballistic test of multilayered armor with intermediate epoxy composite reinforced with jute fabric. Mater Res 18(Supl 2):170–177
7. Luz FS, Monteiro SN, Lima ES, Lima EP Jr (2017) Ballistic application of coir fiber reinforced epoxy composite in multilayered armor. Mater Res 20(Suppl. 2):23–28. https://doi.org/10.1590/1980-5373-MR-2016-0951
8. Monteiro SN, Candido VS, Braga FO, Bolzan LT, Weber RP, Drelich JW (2016) Sugarcane bagasse waste in composites for multilayered armor. Eur Polym J 78:173–185
9. Monteiro SN, Assis FS, Ferreira CL, Simonassi NT, Weber RP, Oliveira MS, Pereira AC (2018) Fique fabric: a promising reinforcement for polymer composites. Polymers 10(246):10. https://doi.org/10.3390/polym10030246
10. Monteiro SN, Louro LH, Trindade W, Elias CN, Ferreira CL, Lima ED, Lima É Jr. (2015) Natural curua fiber-reinforced composites in multilayered ballistic armor. Metall Mater Trans A 46:4567
11. Monteiro SN, Milanezi TL, Louro LH, Lima EP Jr, Braga FO, Gomes AV, Drelich JW (2016) Novel ballistic ramie fabric composite competing with kevlar fabric in multilayered armor. Mater Design 96:263–269
12. Monteiro SN, Pereira AC, Ferreira CL, Lima EP Jr, Weber RP, Assis FS (2018) Performance of plain woven jute fabric-reinforced polyester matrix composite in mutilayered ballistic system. Polymers 10(3):230
13. Garcia Filho FC, Monteiro SN (2018) Piassava fiber as an epoxy matrix composite reinforcement for ballistic armor applications. JOM. https://doi.org/10.1007/s11837-018-3148-x
14. Nascimento LF, Louro LH, Monteiro SN, Gomes AV, Lima Jr. EP, Marçal RL (2017) Ballistic performance in multilayer armor with epoxy composite reinforced with malva fibers. In: Proceedings of the 3rd Pan American materials congress. The minerals, metals & materials society, pp 331–338. https://doi.org/10.1007/978-3-319-52132-9_33
15. Nascimento LF, Louro LH, Monteiro SN, Lima Jr. EP, Luz FS (2017) Mallow fiber-reinforced epoxy composites in multilayered armor for personal ballistic protection. JOM. https://doi.org/10.1007/s11837-017-2495-3
16. Rohen LA, Margem FM, Monteiro SN, Vieira CM, Araujo BM, Lima ES (2015) Ballistic efficiency of an individual epoxy composite reinforced with sisal fibers in multilayered armor. Mater Res 18(Suppl 2):55–62

17. Santos J, Marçal R, Jesus P, Gomes A, Lima Jr. EP, Monteiro SN, Louro L (2017) Effect of LiF as sintering agent on the densification and phase formation in Al_2O_3-4 Wt Pct Nb_2O_5 ceramic compound. Metall Mater Trans A. https://doi.org/10.1007/s11661-017-4271-y
18. Monteiro SN, Lopes FP, Barbosa AP, Bevitori AB, Da Silva IL, Da Costa LL (2011) Natural lignocellulosic fiber as engineering materials-an overview. Metall Mater Trans A, 2963–2974
19. NIJ Standard - 0101.04 (2000) Ballistic resistance of law enforcement and corrections standards and testing program, U.S. Department of Justice, Office of Justice Programs, National Institute of Justice

Part VI
Poster Session

Charpy Impact Test of Polymer Composites with Epoxy Resin Reinforced Jute Fabric

José Gustavo de Almeida Machado, Juliana Peixoto Rufino Gazem de Carvalho, Anna Carolina Cerqueira Neves, Felipe Perissé Duarte Lopes, Sérgio Neves Monteiro and Carlos Mauricio Fontes Vieira

Abstract Nowadays, there is a growing need for the use of environmentally sustainable materials what impulse the research to find eco-friendly materials. The use of natural vegetable fibers has been studied and improved by the market demands by replacing the more expensive and polluting synthetic fibers, such as the glass fiber. The aim of the present work is to investigate the properties of jute fabric as reinforcement of epoxy matrix composites. The test showed that the use of the jute fabric as reinforcement generates a significant improvement in the mechanical properties evaluated in this present work.

Keywords Natural fiber · Jute fabric · Epoxy resin · Impact test

Introduction

One of the major concerns of modern society is the environmental impacts caused by industrial activities, generating the interest of researchers in discovering new materials that can be classified as eco-friendly materials. In this context, one of the possible solutions is the use of natural fibers to replace synthetic ones. Synthetic fibers are non-recyclable and have high production cost and high energy consumption compared to natural fibers that are abundant, renewable, recyclable, biodegradable, and essentially carbon neutral [1–4]. The economic advantage of natural fibers is the considerably lower price compared, for example, to glass fibers, which is the cheapest among synthetic fibers.

J. G. de Almeida Machado · J. P. R. G. de Carvalho (✉) · A. C. C. Neves · F. P. D. Lopes · C. M. Fontes Vieira
State University of the Northern Rio de Janeiro - UENF, Advanced Materials Laboratory - LAMAV, Avenida Alberto Lamego, 2000, 28013-602 Campos Dos Goytacazes, Brazil
e-mail: julianarufino.arq@gmail.com

S. N. Monteiro
Military Institute of Engineering - IME, Materials Science Department, Praça General Tibúrcio, 80, 22290-270 Urca, Rio de Janeiro, RJ, Brazil

S. Ikhmayies et al. (eds.), *Green Materials Engineering*, The Minerals, Metals & Materials Series, https://doi.org/10.1007/978-3-030-10383-5_23

In addition, the natural fibers are capable to replace the synthetic fibers used in some automotive and aeronautical components, due to good toughness and less abrasion to equipment used in composite processing [1–5].

This study aims to analyze the properties of jute fabric as reinforcement of polymeric matrix composites and its behavior regarding the impact resistance evaluated by the Charpy impact test.

Methodology

The 5 m of jute fabric was purchased in a store in the city of Campos dos Goytacazes (RJ). The jute fabric was cut into rectangles of 155 × 125 mm, washed in running water and placed in the oven for 24 h at a temperature of 60 °C until its mass was constant, and the fabric was completely dry.

To determine the length and spacing values between the fibers, three points of each fabric layer were measured, and a total of 20 fabric layers were used to obtain the mean values. The equipment used to perform and determinate the measurement was the PANTEC profile projector. The measurements were performed at the Laboratory of Advanced Materials (LAMAV), located at UENF. The average value of the fiber spacing length was (2.664 ± 0.128) mm, and the mean value of the fiber spacing width was (3.109 ± 0.123) mm.

Determination of the Fiber Density

The density calculation in relation to the distilled water was carried out using a pycnometer and a digital scale with an accuracy of ±0.001 g. First, the tare of the digital scale was made with the empty pycnometer, and then the mass of the pycnometer + water was weighed. Soon after, the mass of the pycnometer/solid/distilled water set was measured, thus obtaining the apparent density of the fiber in relation to the water.

Before placing the tissue in the pycnometer with distilled water, it was kept under vacuum for 2 h in order to remove the existing bubbles.

Parallel to the test, the moisture correction factor that could exist in the material was calculated. For this, a quantity of tissue was weighed and placed in a greenhouse for 24 h, and then cooling was done under vacuum in order to avoid the absorption of moisture from the medium.

The expression for calculating the density (ds) of the tissue as a function of the water density was given by Eqs. 1 and 2, where m1 is the mass of the empty pycnometer; m2 is the mass of the pycnometer with the fiber; m3 is mass of the pycnometer with fiber + water; m4 is mass of the pycnometer with water:

$$ds = \frac{\rho s}{\rho\,\text{water}} = \frac{m2 - m1}{(m4 - m1) - (m3 - m2)} \quad (1)$$

$$\rho s = ds \cdot \rho\,\text{water} \quad (2)$$

As the scale was calibrated with the pycnometer, the calculation was performed considering that the mass of the empty pycnometer was zero. To raise the density, a 50-ml pycnometer was used.

Composites

The epoxy resin SQ2001 and hardener SQ3154 from Silaex Química Ltda. were used as the polymer matrix.

The amount of jute fabric layers constituting the samples ranged from 0 to 6 layers. Except the specimen without the jute fabric, the composites were put under 2 MPa pressure in a hydraulic press for 24 h, and after that, the composite was removed from the metallic matrix and left to complete the cure process at room temperature for 7 days. After this period, the samples were placed in the oven for 24 h at 60 °C. The polymeric sample of epoxy resin, without a fabric layer, remained in the mold without the addition of pressure for 24 h. It is important to mention that the ratio of the epoxy resin to the hardener is 2:1.

For each composite plate, at least 10 specimens were cut for the Charpy impact test, dimensions 125 × 12.7 × 10 mm, according to ASTM D6110.

Results and Discussion

The density of the jute fiber obtained by pycnometry is 0.67 g/cm^3. This value was used to prepare the test bodies for the Charpy impact test. Table 1 shows the densities of other fibers compared to the density of the jute fabric [7]. In this way, the jute fabric, in relation to the other fabrics, is considerably light.

Table 2 shows the energy absorbed from the Charpy impact test of the epoxy resin matrix composites reinforced with layers of jute fabrics from 0 to 6 layers. The increase of the number of jute fabric layers in the polymer matrix increased the tenacity of the notch, as expected [8]. The values obtained corroborate with different fabrics, such as cotton fabric with polystyrene matrix [9] and epoxy composites reinforced with malva fibers [10], but toughness was lower in comparison with the polyurethane matrix reinforced with sisal fabric [11].

Figure 1 shows the impact energy change related to the number of strings. In the interval between 3 and 4 layers of jute fabric, there was a significant variation in impact energy. This increase in notch toughness can be explained by the large energy consumed to break the fibers within the Charpy assay procedure [6].

Table 1 Fibers density

Fiber	Density (g/cm³)
Jute	0.67
Cotton	1.5–1.6
Linen	1.5
Sisal	1.5
Carbon	1.4
Aramida	1.4
E-glass	2.5
S-glass	2.5

Table 2 Energy absorbed in Charpy impact test

Quantity of layers	Energy absorbed (J/m)
0	52.15
1	60.35
2	58.42
3	67.04
4	128.42
5	138.70
6	135.79

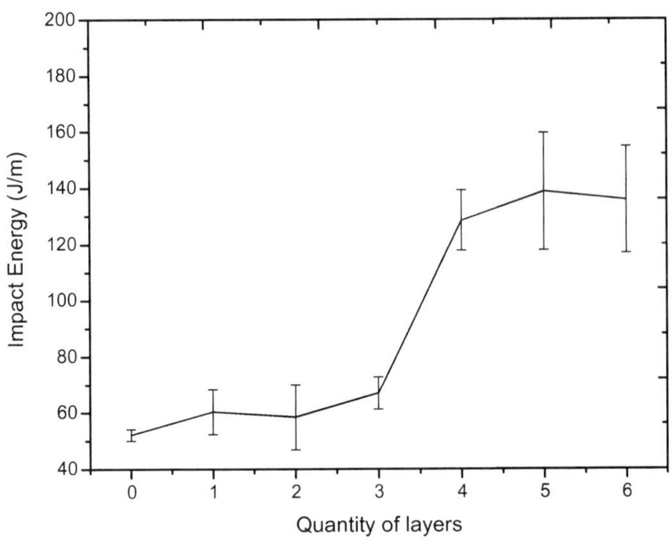

Fig. 1 Variation of impact energy per quantity of fabric layers

Conclusion

The incorporation of the jute fabric as reinforcement of epoxy matrix composited showed a considerable increase in notch toughness as measured by the Charpy assay compared to the properties of pure epoxy matrix samples.

The significantly higher increase in energy occurs when the number of jute fabric layers varies from 3 layers to 4 layers, from 67.04 to 128.42 J/m of energy.

The 1, 2, and 3 layer samples did not show a significant increase of the tenacity to the notch compared to the pure resin. Although with 4, 5, and 6 layers of fabric incorporated, the increase was considerable compared to pure resin.

The specimens with 2 and 6 layers of fabric did not follow the linear behavior observed in certain bands according to the increase of the number of jute fabric layers, with values of 58.42 J/m and 135.79 J/m, respectively. This is attributed to the misalignment of the fibers inside the matrix, reducing the efficiency of the absorption of the energy of the same.

Acknowledgements The authors are grateful for the support of the Laboratory of Advanced Materials—LAMAV/UENF, mainly to Renan and soils laboratory (LSOL), located in CCTA/UENF.

References

1. Nabi Sahed D, Jog JP (1999) Natural fiber polymer composites: a review. Adv Polym Technol 18:221–274
2. Mohanty AK, Misra M, Hinrichsen G (2000) Biofibers, biodegradable polymers and biocomposites: an overview. Macromol Mat Eng 276(277):1–24
3. Monteiro SN, Lopes FPD, Ferreira AS, Nascimento DCO (2009) Natural fiber polymer matrix composites: cheaper, tougher and environmentally friendly. JOM 61(1):17–22
4. Bledzki AK, Gassan J (1999) Composites reinforced with cellulose-based fibers. Prog Polym Sci 4:221–274
5. Gore A (2006) An inconvenient truth. The planetary emergency of global warming and what we can do about it. Rodale Press, Emmaus, Pennsylvania, USA
6. Camerini AL (2008) Caracterização e propriedades de compósitos de tecido de juta reforçando matriz de polietileno reciclado. UENF
7. Gon D, Das K, Paul P, Maity S (2012) Jute composites as wood substitute. Int J Text Sci
8. Callister Jr. WD (2000) Materials science and engineering—An introduction, 5a Edição. Wiley, New York
9. Borsoi C, Scienza LC, Zaltera AJ, Angrizani CC (2011) Obtenção e Caracterização de Compósitos Utilizando Poliestireno como Matriz e Resíduos de Fibras de Algodão da Indústria Têxtil como Reforço, Polímeros: Ciência e Tecnologia, vol 21, núm 4. Associação Brasileira de Polímeros, São Paulo, Brasil, pp 271–279
10. Margem JI (2013) Estudo das Características Estruturais e Propriedades de Compósitos Poliméricos Reforçados com Fibras de Malva. UENF
11. Silva RV (2003) Compósitos de Resina Poliuretano derivada de Óleo de mamona e fibras vegetais. USP

Development of Silicate Glasses with Granite Waste

Michelle Pereira Babisk, Vinicius Rodrigues Gomes, Juraci Aparecido Sampaio, Monica Castoldi Borlini Gadioli, Francisco Wilson Hollanda Vidal and Carlos Mauricio Fontes Vieira

Abstract Granite is an igneous stone with high quartz (SiO_2), whose processing generates millions of tons of fine residue annually in Brazil. Among the various definitions of glass, the most widely used is that it is a physically homogeneous substance obtained by cooling a melting inorganic mass, which solidifies without crystallizing, therefore, glasses do not have a regular atomic arrangement and, hence, are called amorphous. The objective of this work was to characterize a granite chemically and morphologically and to use it as the main raw material for producing silicate glasses of the soda-lime and borosilicate types. The characterization of the granite waste revealed that the silica (SiO_2) is the major component, followed by the alumina (Al_2O_3), in grains of angular shapes, which favors the glass production process. The glasses produced were colored (green and amber), totally amorphous and with densities consistent with the values quoted in the literature, for each type of glass. The results show that this waste can be used as raw material in glassmaking, thus obtaining a correct destination for this waste.

M. P. Babisk (✉) · V. R. Gomes · C. M. F. Vieira
Laboratório de Materiais Avançados, Universidade Estadual do Norte Fluminense Darcy Ribeiro, Campos dos Goytacazes, RJ, Brazil
e-mail: michellebabisk@hotmail.com

V. R. Gomes
e-mail: vinicius.rod3@gmail.com

C. M. F. Vieira
e-mail: vieira@uenf.br

J. A. Sampaio
Laboratório de Ciências Físicas, Universidade Estadual do Norte Fluminense Darcy Ribeiro, Campos dos Goytacazes, RJ, Brazil
e-mail: jsampaio@uenf.br

M. C. B. Gadioli · F. W. H. Vidal
Núcleo Regional do Espírito Santo, Centro de Tecnologia Mineral, Cachoeiro de Itapemirim, ES, Brazil
e-mail: monicaborlini@gmail.com

F. W. H. Vidal
e-mail: fwhollanda@gmail.com

S. Ikhmayies et al. (eds.), *Green Materials Engineering*, The Minerals, Metals & Materials Series, https://doi.org/10.1007/978-3-030-10383-5_24

Keywords Granite waste · Silicate glasses · Soda-lime glass · Borosilicate glass

Introduction

Stones are generally defined as natural solid bodies, formed by aggregates of one or more crystalline minerals. From the commercial point of view, ornamental and cladding stones are basically subdivided into two large groups, granites and marbles. As granites, they are generally made up of silica stones, while marbles, carbonate stones [1].

In the case of the production of ornamental stones, among other steps, the sawing of blocks is made to turn them into semifinished plates. The equipment used for the sawing process is called the loom, which consists of steel blades or diamond wire blades. In this phase, expressive amounts of waste are generated, according to surveys carried out by Vidal et al. (2013), the volume of fines generated, which corresponds to the volume of the cutting furrows, is equivalent, on average, to 26% of the initial volume of the block. In Brazil, it is estimated that approximately 2 Mt of fine waste is generated annually [2, 3].

According to Abirochas (2018) granites made up more than 87% of Brazilian stone exports in 2017. Granite is an igneous stone, consisting mainly of feldspar, quartz, minerals with iron. The quartz has a crystalline structure composed of silica tetrahedra, silicon dioxide (SiO_2), and feldspar, which responds to the highest percentage concentration in the granite stone formation, have a crystallized silica structure. Feldspars are a series of aluminum silicates and alkaline or earth alkaline bases, both potassium and sodium feldspars are generally present. The micas are the minerals that present in smaller percentage in the composition belong to the group of aluminum silicates and other metals [4, 5].

Silica is a naturally occurring glass-forming oxide, and the addition of oxides imparts distinct characteristics. These oxides added to the silica may be either network modifying oxides or intermediate oxides. The network modifying oxides break the glass structure formed by silica, reducing the melting point and the viscosity of the fluid, and the intermediaries may act as modifiers or as network formers [6].

Since Zachariasen (1932) published and based his paper "The Atomic Arrangement in Glass" on X-ray diffraction results showing that the glasses differ from the crystals because they do not have a long-range symmetrical and periodic network except for their basic unity, different definitions of glass have appeared in the scientific literature, but several authors have maintained this dependence in their definitions [7–12].

ASTM (American Society for Testing and Materials) defines glass as an inorganic melt that has been cooled to a rigid condition without crystallization, and is therefore referred to as amorphous materials. A material is amorphous when it does not present a long-range order in the atomic arrangement, that is, when there is no regularity of its molecular constituents on a scale larger than a few times the size of these groups [6].

Industrially, the concept of glass can be restricted to the products resulting from the fusing of oxides or their derivatives and mixtures, generally having as main constituent the silica, for the common glasses. Within this scenario and context, the objective of this work is to characterize a fine granite waste and to use it as the main raw material to produce silicate glasses of the soda-lime and borosilicate types.

Experimental

In this work, we have used granite waste as silica source and commercial oxides to produce soda-lime and borosilicate glasses.

The granite residue was collected in a sawmill in Cachoeiro de Itapemirim/ES, Brazil. First, the waste was air dried to remove excessive moisture followed by later drying in an oven at ±110 °C for 24 h. The batch was disaggregated with mortar and pestle and sieved in 40 mesh.

For the addition of the other component oxides of the glasses, the following chemical reagents type A.P.: sodium (Na_2CO_3) and calcium ($CaCO_3$) carbonates, alumina (Al_2O_3) and boric acid (H_3BO_3) were used.

The characterization of the waste was done by chemical analysis by X-ray fluorescence (XRF) and morphological by scanning electron microscopy (SEM). The XRF was performed on a Panalytical Axios spectrometer (WDS-2), and the SEM micrographs were obtained on a MEV FEV Quanta 400 scanning electron microscope.

The glass compositions (Table 1) were based on a glass used in beverage packaging for soda-lime and Pyrex® glasses for borosilicate [6]. They were cast in a furnace Sentro Tech Corp. ST-1700C-445 at 1500 °C. The samples were subsequently annealed at 600 °C for 2 h.

The X-ray diffraction patterns of the glass samples were collected in a Bruker-D4 Endeavor equipment, under the following operating conditions: Co Ka (35 kV/40 mA); goniometer velocity of 0.02° 2θ per step with counting time of 1 s per step and collected from 5 to 80° 2θ. The glass samples densities were obtained by the Archimedes method at 22 °C using water as immersion fluid.

Table 1 Composition of glasses wt%)

	SiO_2	Al_2O_3	Na_2O	CaO	B_2O_3
Soda lime	72	2	13.5	12.5	–
Borosilicate	79	2	6	–	13

Results and Discussion

The main raw material for manufacturing silicate glasses is sand, which has the function of providing silica (SiO_2) to the glass [13]. As this research aimed to use the fine granite waste to replace the sand, the discussion of its characterization will be made in comparison of this raw material.

Industrially, the angular grains of the sand favor the process of production of the glass, because the fusion begins at the tips and edges of the grains. In the micrographs obtained by MEV of the fine quartzite residue, shown in Fig. 1, the morphology of the particles of the residue can be observed and that these have angular grains (Fig. 1b) [14].

In a non-focused heating on crystals, it is common for the melt to occur on the surface, since the melting kinetics favors the ends of the crystals, so fusion occurs "from the outside to the inside." Quartz, for example, can remain solid up to 300 °C above its natural melting temperature (1400 °C) in the regions of the center of the material because of this effect of melting kinetics [15].

Table 2 shows the chemical composition of the granite waste. It is possible to observe the majority presence of SiO_2 (69.9%), glass-forming oxide, which confirms its efficiency as a silica supplier for the production of silicate glasses, followed by Al_2O_3 (17%) intermediary oxide. Significant amounts of CaO, K_2O and Na_2O, network modifying oxides have also been identified. Content of 1.3% of Fe_2O_3 was identified; in this work, this oxide will play a fundamental role as it will act as a dye in the formulation of the glasses.

Fig. 1 Micrographs of fine granite waste: **a** 1000× and **b** 2000×

Table 2 Chemical composition of the granite waste (% weigth)

Components	SiO_2	CaO	K_2O	Na_2O	Al_2O_3	Fe_2O_3	MgO	P_2O_5	TiO_2	LoI
	69.9	1.5	3.4	5.5	17	1.3	0.46	0.14	0.24	0.55

LoI: Low loss on ignition

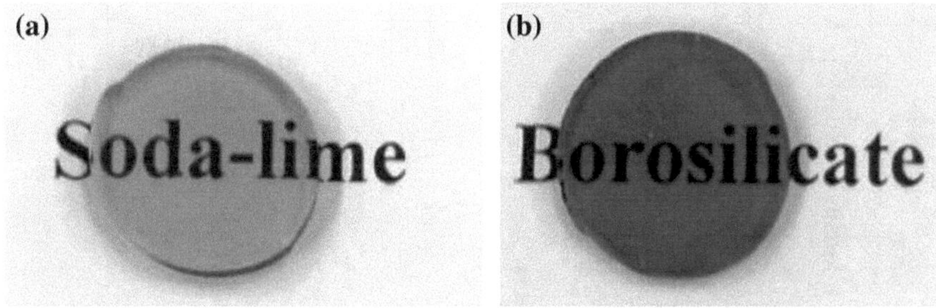

Fig. 2 Pictures of the glasses produced with granite waste: **a** Soda lime and **b** Borosilicate

Table 3 Density of glasses produced with granite waste (g/cm^3)

Glasses	Soda lime	Borosilicate
Density	2.55	2.37

Due to the presence of iron oxide in the granite waste, the soda-lime glass showed a greenish color, while the borosilicate showed amber coloration.

In commercial terms, the color of glass can be very important in marketing matters as it can help in choosing the product. Besides the esthetic function, the color of the glass also has a utilitarian function. Depending on the element, it can filter the light, letting some rays pass and retaining others. That is why green and amber bottles are used for drinks and medicines, because these glasses prevent the passage of certain (ultraviolet) radiations, which deteriorate the products.

Figure 2 shows the images of the glasses produced where it is possible to visualize the text through the samples with approximately 2 mm of thickness, it can be observed that both the soda-lime glass and the borosilicate have good transmittance in the visible region.

The X-ray diffractograms of the glasses produced are shown in Fig. 3. It can be observed the absence of crystalline peaks, and the spectra represent a typical amorphous band around 27°, due to the majority presence of silica in the glasses. This result characterizes the glasses as amorphous, they do not present a long-range order in the atomic arrangement.

The densities of the glasses produced are presented in Table 3 and are close to the values in the literature. The densities of the soda-lime glasses are compatible with the increased density that alkali incorporation provides in the glass silica network, which leaves densities around 2.57 g/cm^3, very close to the value found for the glass produced with the granite waste. The packing of the SiO_2 and B_2O_3 networks is similar, which maintains a similar oxygen atom packing and there is no significant variation in density between the glassy silica and the borosilicate glass, keeping the density in values around 2.22 g/cm^3, the presence of the alkali oxides in the waste, thus adding more of these in the composition, may have influenced and a small increase in the density value of the borosilicate glass produced with the waste can be observed [16].

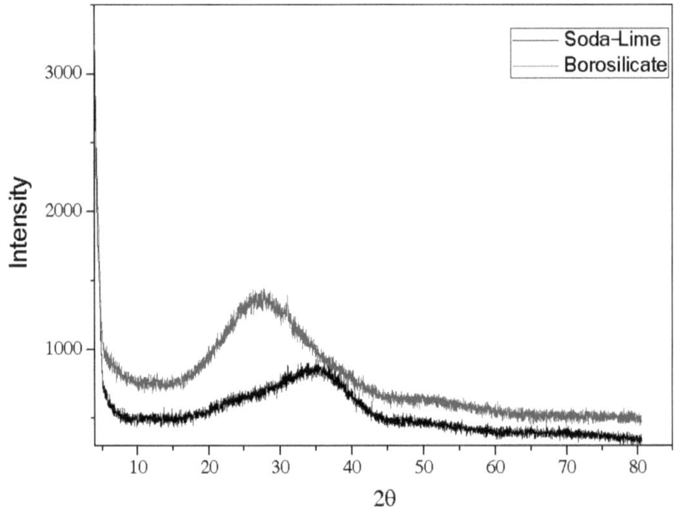

Fig. 3 X-ray diffractograms of the glasses produced with granite waste

Conclusions

It is justified the use of granite waste for the production of silicate glass, soda-lime and borosilicate types, its characterization proved the majority composition of silica (69.9%) and alumina (17%), as well as oxides alkaline and alkaline earths used in the composition of the glasses produced. The morphology of the waste in angular grains helps the fusion and, consequently, the productive process. The glasses produced were totally amorphous and colored (green and amber), due to the presence of iron oxide in the granite waste, the color of the glasses can be important in esthetic matters, and also useful, because these glasses block certain radiations. Despite the color, they presented good transmittance in the region of the visible and densities compatible with the respective types of glasses. The results show that this waste can be used as raw material in the manufacture of silicate glass, thus obtaining a correct destination for this waste generated in the order of millions of tons.

Acknowledgements The authors thank the Coordenação de Aperfeiçoamento de Pessoal de Nível Superior, CAPES, program PNPD20131134 - 31033016005P8 UENF/Materials Engineering and Science, and the Conselho Nacional de Desenvolvimento Científico e Tecnológico, CNPq, process 162622/2017-1 for the support of the research project.

References

1. Chiodi Filho C, Rodrigues EP (2009) Guide to application of rocks in coatings. Bula Project-Abirochas (in portuguese)
2. Calmon JL, Silva SAC (2006) Marble and granite in espírito santo: environmental problems and solutions. In: Domingues AF, Boson PHG, Alípaz S (eds) Water resources management and mining. National Water Agency-ANA, Brazilian Institute of Mining-IBRAM, Brasília, pp 199–231 (in portuguese)
3. Vidal FWH, Azevedo HCA, Castro NF (eds) (2013) Ornamental stone technology: research, mining and processing. CETEM/MCTI, Rio de Janeiro (in portuguese)
4. Abirochas, In: Balance of Brazilian exports and imports of ornamental stones in 2017. Report 01/2018. http://abirochas.com.br/wp-content/uploads/2018/03/Informe-01-2018-Balanco-2017.pdf. Accessed 12 June 2018 (in portuguese)
5. Andrade MC (2006) Characterization and utilization of feldspar contained in fine granite quarries. Masters Degree dissertation, Universidade Federal do Rio de Janeiro (in portuguese)
6. Akerman M (2000) Introduction to glass and its production. Glass School-ABIVIDROS. http://www.certev.ufscar.br/documentos/arquivos/introducao-ao-vidro. Accessed 12 June 2018 (in portuguese)
7. Zachariasen WH (1932) The atomic arrangement in glass. J Am Chem Soc
8. Zarzycki J (1991) Glasses and amorphous materials. Mater Sci Technol, Wiley
9. Varshneya AK (1993) Fundamentals of inorganic glasses. Academic Press, London
10. Doremus RH (ed) (1994) Glass science. Wiley Interscience, New Jersey
11. Shelby JE (1997) Introduction to glass science and technology. Royal Society of Chemistry, London
12. Zanotto E, Mauro J (2017) The glassy state of matter: its definition and ultimate fate. JNCS 471:490–495
13. Luz AB, Lins FF (2005) Industrial sand. In: Industrial rocks and minerals 2005. CETEM, Rio de Janeiro, pp 107–126 (in portuguese)
14. Nava N (1997) Geology of industrial sands. In: Major Brazilian mineral deposits 2005. CETEM, Rio de Janeiro, pp 325–331 (in portuguese)
15. Ainsile NG, Mackenzie JD, Turnbull D (1961) Kinetic of fusion of quartz and cristobalite. JPC 65:1718–1724 (in portuguese)
16. Scholze H (1991) Glass—Nature, structure and properties. Springer, New York

Evaluation of Feldspathic Rock Waste on the Production of Structural Ceramics with Greater Value Added

L. F. Amaral, M. Nicolite, G. C. G. Delaqua, S. N. Monteiro and Carlos Mauricio Fontes Vieira

Abstract Santo Antônio de Pádua, state of Rio de Janeiro, is considered one of the municipalities with the highest production of ornamental rocks in the state of Rio de Janeiro. The rocks found in this region are classified as metamorphic of the mylonite gneiss type. The residue used in this work comes from the rock commonly known as Carijó, rich in potassium feldspar and thus having a light color. In this way, this work aims to investigate the influence of this residue incorporation on the coloration of the ceramic products, as well as its technological properties, considering that the pieces of light color have higher added commercial value. Therefore, 13 specimens were prepared per batch, with three different percentages of residues incorporated, being 0, 20, and 40%, and three different levels of sintering temperature, being 800, 900, and 1000 °C. Among the analyzed properties, it was observed a tendency of water absorption reduction, mechanical resistance increasing, reduction of the firing linear shrinkage, apparent dry density increasing, and improvement of the plasticity with the incorporation of the residue was observed.

Keywords Clay · Ornamental rock · Waste incorporation · Gnaisse

L. F. Amaral (✉) · M. Nicolite · G. C. G. Delaqua · C. M. Fontes Vieira
Advanced Materials Laboratory – Science and Technology Center, State University of North Fluminense Darcy Ribeiro, Alberto Lamego Avenue, 2000, Campos dos Goytacazes, RJ 28013-602, Brazil
e-mail: lfamaral@pq.uenf.br

M. Nicolite
e-mail: geovanagirondi@gmail.com

G. C. G. Delaqua
e-mail: micaela_uenf@yahoo.com.br

C. M. Fontes Vieira
e-mail: carlos.vieira@pq.cnpq.br

S. N. Monteiro
IME (Military Institute of Engineering), Praça General Tibúrcio 80, Praia Vermelha, Urca, Rio de Janeiro, RJ 22290-270, Brazil
e-mail: snevesmonteiro@gmail.com

S. Ikhmayies et al. (eds.), *Green Materials Engineering*, The Minerals, Metals & Materials Series, https://doi.org/10.1007/978-3-030-10383-5_25

Introduction

In the municipality of Santo Antônio de Pádua, located 150 km from Campos dos Goytacazes—RJ, there is intense industrial activity related to the extraction and processing of ornamental rocks. After mining, the ornamental rocks are submitted to sawing to obtain blocks, during this operation is generated the sludge composed by water and fine particles of rock, and this sludge is commonly referred to as rock waste.

The waste generated in this process has already been widely investigated as a possible raw material in civil construction with the advantage of improving the technological properties of mortars and ceramic blocks [1–4]. The ceramic industry has stood out in recent years as an alternative to the destination of large quantities of solid waste, thus contributing to environmental preservation [5]. A rock residue called "white wood stone" or "Carijó," generated in Santo Antônio de Padua's ornamental rock processing industries, has been attracting attention because of its white coloration.

It is general knowledge accepted that ceramic products that have a light color after firing have a higher economic value added, according to consumer preference. In addition, the final disposal of this waste has brought serious environmental problems. Since most industries do not have adequate sludge treatment, this waste is polluting the soil and underground waters as well as obstructing rivers and lakes. Thus, this work investigated the effects of the incorporation of "white stone" rock residue on the technological properties of traditional ceramic products, since its blending with clays after firing provides the clear color required by the consumer.

Methodology

The basic raw materials used in this investigation were two different local clays from Campos dos Goytacazes, as well as rock waste rich in potassium feldspar, from Santo Antônio de Pádua. Upon receipt, the raw materials were dried at 110 °C, manually disintegrated with a crusher, and then screened at 20 mesh (0.840 mm).

Three experimental bodies, in wt%, were formulated by incorporating 20 and 40% of rock waste in the clayey body. The plasticity of the ceramic bodies was evaluated by the Atterberg methods [6, 7].

The raw materials were initially characterized in terms of its chemical composition. The chemical composition was determined by X-ray fluorescence (XRF) in a Philips PW 2400 equipment. The particle size distribution was determined by both, sieving and sedimentation methods, following the norm [8].

In order to determine the technological properties of the ceramic bodies, 11.5 cm rectangular specimens with 1.1 cm × 2.54 cm in cross section were shaped by uniaxial pressing. Initially, the specimens were dried at 110 °C until constant weight. Finally, the specimens were sintered at temperatures of 800, 900, and 1000 °C in an electric laboratory kiln with a 180 min socket at maximum temperature, using a heating/cooling rate of 2 °C/min. The measured technological properties were: dry apparent density, measured by the dimensional method dividing the dry weight by the external volume; firing linear shrinkage; water absorption and mechanical strength, obtained by the flexural rupture strength, using the three points method.

Results and Discussion

Table 1 shows the chemical composition of the raw materials. It can be noted for clayey body a relatively low relation for SiO_2/Al_2O_3 ratio, being close to 1.17, that is the theoretical ratio value for kaolinite. This indicates an elevated fraction of clay minerals. Moreover, the low amount of Fe_2O_3 is sufficient to produce light colored traditional ceramic products. It also should be noted that the clayey body has relatively low content of alkaline oxide (K_2O) and earth alkaline oxides (CaO), which act as fluxes to improve the sintering mechanisms. The high loss on ignition is proportional to the clay mineral content, so the higher the content of this mineral, the greater is the loss of water caused by its dehydroxylation during sintering.

The chemical composition of rock waste is also presented in Table 1. In addition to SiO_2 and Al_2O_3, the presence of alkaline oxides is also prevalent in this raw material; thus, this waste could act as fluxing agent during sintering step at traditional ceramic product.

Table 2 contains the granulometric distribution of the clay and the residue. The different grain diameter ranges were classified according to grain size, clay, and silt, as Santos [9] suggests through the International Society of Soil Science. According to this classification, the argillomineral fraction corresponds to the particles with equivalent spherical diameter (deq) <2 μm, the silt fraction corresponds to the particles with deq between 2–20 μm and the sand fraction corresponds to the particles with deq >20 μm. As can be seen, the clay has a granulometric distribution rich in particles related to clay minerals. This characteristic is undesirable for the production of red ceramics, due to the negative influence of the excessive presence of clay minerals. This influence is related to some difficulties in the production process, such as the greater need of water for product conformation and greater amount of energy consumed during the drying and also the burning phase. In this way, it is possible to

Table 1 Raw materials chemical composition

	Al_2O_3	SiO_2	K_2O	Fe_2O_3	CaO	TiO_2	SO_3	Other oxides	LoI[a]
Clayey body	36.62	46.12	0.08	1.67	0.13	1.71	–	0.33	13.34
Waste	17.91	67.25	8.94	1.28	2.69	0.40	1.30	0.23	–

[a]Loss on fire

Table 2 Raw materials granulometric distribution

Raw materials	Granulometric distribution		
	<2 μm	2–20 μm	>20 μm
Clayey body	87%	6%	7%
Rock waste	5%	50%	45%

Table 3 Apparent dry density (D_{ap})

	Clayey body		
	0%	20%	40%
D_{ap} (g/cm^3)	1.67	1.74	1.72

reduce the relative amount of clay minerals present in the ceramic mass, in order to improve the processing, as well as the properties of the final product, according to the granulometric distribution of the residue.

The values for dry bulk density of the specimens are shown in Table 3. It is possible to notice that the incorporation of residue provides the increase of the density, indicating that there is the optimization of the granulometric distribution. This is due to the different grain sizes of each raw material, in which its mixture synergistically converts to increase dry density.

Figure 1 shows the extrusion prognostic [10] with the location of the investigated formulations. This figure indicates the Atterberg limits for clayey bodies based on the combined positions of the plastic limit and the plastic index. The indicated rectangular areas in the figure can be associated with optimum extrusion, acceptable extrusion, and non-recommended extrusion. In this figure, it is worth observing that the typical clayey body (CB), owing to it excessive plasticity due to the high amount of clay minerals [11], is located in the non-recommended extrusion area. It is also observed that the formulation CB (20%), that contains 20 wt% of waste, decreased the plastic limit and practically maintained the plastic index in comparison with the CB, changing the area of acceptable extrusion. The CB (40%) was the only one able to locate at the optimal extrusion area.

Figure 2 shows the mean values and standard deviation of the water absorption of all sample groups. It was verified that the incorporation of the residue was able to reduce the values of the water absorption for all temperatures, mainly when incorporating the percentage of 20%, which reached the lower levels of water absorption. The reduction of water absorption is mainly due to the reduction of the open porosity provided by the increased apparent dry density. In addition, the presence of alkaline oxides in the residue assists in the formation of viscous flow through the eutectic reaction with silicon, which is the main sintering mechanism for the traditional ceramics industry.

It is also possible to observe the high water absorption for the clayey body without residue at temperatures of 800 and 900 °C. This behavior is associated with a refractory behavior of a kaolinite clayey body, with low amounts of flux oxides and high amount of alumina, that difficult the closure of porosity. At temperatures close to 1000 °C, it may be observed an abrupt decrease in water absorption and increase in

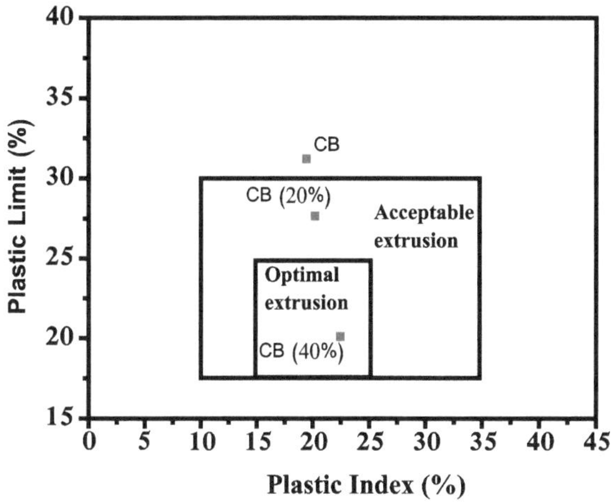

Fig. 1 Extrusion prognosis through the plastic limit and plastic index

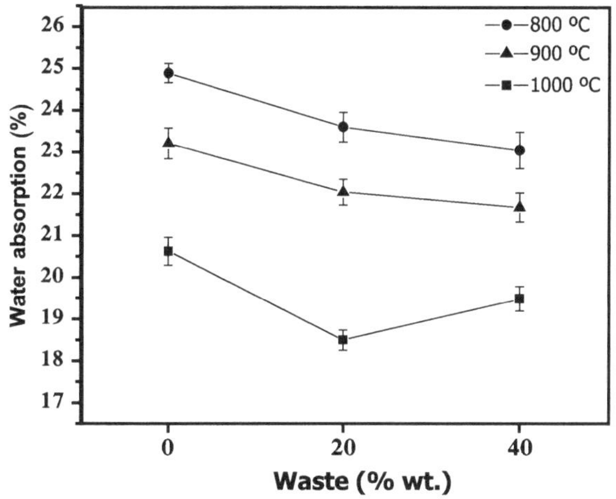

Fig. 2 Water absorption as function of waste incorporation

linear shrinkage (Fig. 3). This behavior is associated with the sintering mechanisms, diffusion in the solid state and liquid phase formation, which act effectively in the Campos dos Goytacazes clays from 1000 °C. It is also observed that the level of 20% water absorption, maximum acceptable value for the type Roman roofing tiles, according to technical standard [12], is only reached at 1000 °C for clayey bodies containing 20 and 40% of residue.

The linear shrinkage, Fig. 3, has direct correlation with water absorption. Thus, the higher the linear shrinkage, it can be inferred that there is a higher level of sintering that leads to the reduction of total porosity. However, this correlation is significant

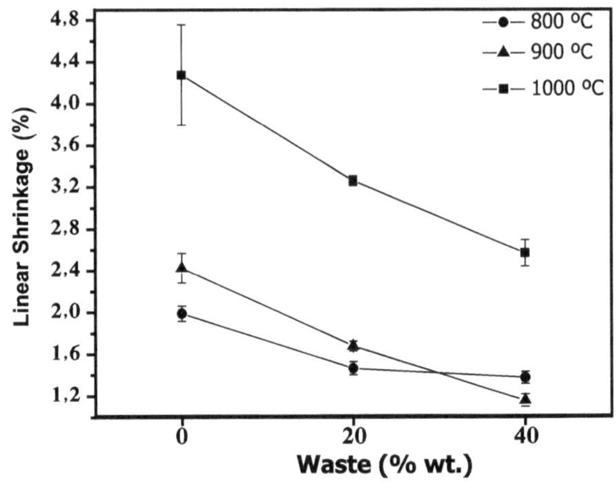

Fig. 3 Linear shrinkage as function of waste incorporation

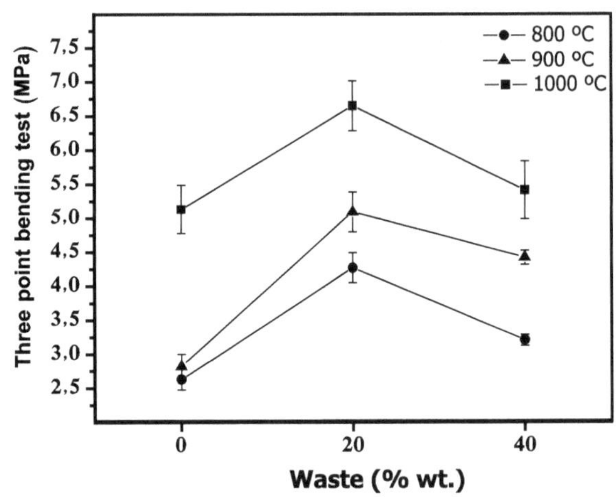

Fig. 4 Flexural rupture strength as function of waste incorporation

only to a certain extent, since the water absorption is related only to the open porosity of the structure.

It is generally known that excessive shrinkage is capable of causing stresses in the ceramic body, and if it does not support the tension exerted, cracking may occur and the product should be discarded. Thus, it is interesting to note that the residue has the ability to reduce the linear shrinkage of the ceramic structure, allowing avoiding product defects.

In Fig. 4, the values for mechanical strength are shown, by means of the three-point bending rupture stress, for all the formulations evaluated. It is possible to notice that the incorporation of the residue to the clayey body helped in the increase of the mechanical resistance for all the evaluated temperatures. This result corroborates with the results of apparent density and water absorption. Thus, the denser the structure, the greater the contact points and consequently the greater the mechanical strength.

At temperatures close to 1000 °C, the formulations present a significant increase in mechanical strength, as a consequence of liquid phase formation with partial dissolution of quartz particles. Finally, it is also observed that the formulation containing 20% of waste was the only that reached 6.5 MPa of flexural rupture strength, that is the minimum recommended value.

Conclusions

This work showed that with the incorporation of 20% of rock residue in the studied clayey body and sintering at temperature of 1000 °C reached the best result of water absorption and mechanical resistance, which are in accordance with the technical standards for roofing tiles. This formulation can also be used for the production of bricks by sintering at 900 °C.

The formulation containing 40% of the residue, although it did not provide the best results in the physical and mechanical properties, provided the best extrusion prognosis, reducing the plasticity of the mass, and it can be used to produce bricks at a temperature of 1000 °C.

Acknowledgements The author thanks to the financial support of Brazilian agencies FAPERJ (n. E-26/202.773/2017) and CNPq (392930/2014-0).

References

1. Moura WA, Gonçalves JP, Leite RS (2002) Use of the cutting residue of marble and granite in coating mortar and floor tiles (in port). Sitientibus 26:46–61
2. Faial ASR, Xavier GC, Alexandre J, Maia PCA, Albuquerque FS (2012) Use of ornamental stone residue in the production of multiple use mortar (in port). In: 56º Congresso Brasileiro de Cerâmica, Curitiba, PR, Brasil
3. Carvalho JPRG (2015) Study of technical parameters of ceramic pavement with incorporation of ornamental rock residue. Dissertation thesis, State University of Northern Rio de Janeiro
4. Vieira CMF, Soares TM, Sánchez R, Monteiro SN (2004) Incorporation of granite waste in red ceramics. Mater Sci Eng 373(1–2):115–121
5. Oliveira SEM, Sampaio VG, Holanda JNF (2006) Incorporação de Resíduo de ETAs em Cerâmica Vermelha. In: 17º Materials Science and Engineering Congress (in port.) 11. Foz do Iguaçu –PR
6. Brazilian Association of Technical Standards (1984) Soil—Plasticity limit determination. NBR 7180, Rio de Janeiro
7. Brazilian Association of Technical Standards (1984) Soil—Liquid limit determination. NBR 6459, Rio de Janeiro
8. Brazilian Association of Technical Standards (1984) Soil—Grain size analysis. NBR 7181, Rio de Janeiro
9. Santos OS (1989) Science and technology of clays (in port). São Paulo, SP: Edgard Blücher

10. Marsigli M, Dondi M (1997) Plasticitá delle argille italiane per laterizi e previsione del loro comportamento in foggiatura. L'Industria dei Laterizi 46:214–222
11. Alexandre J, Azevedo ARG, Xavier GC, Pedroti LG, Vieira CMF, Monteiro SN (2014) Study of a clayey soil used in the fabrication of red ceramics in Campos dos Goytacazes, Brazil. Mater Sci Forum (Online) 798–799:15–20
12. Brazilian Association of Technical Standards (2009) Ceramic components—Ceramic roof tiles-Terminology, requirements and testing methods. NBR 15310, Rio de Janeiro

Izod Impact Testing in Composites with Vegetal Polyurethane Matrix Reinforced by Cotton Fabric

Carolina Gomes Dias Ribeiro, Juliana Peixoto Rufino Gazem de Carvalho, Felipe Perissé Duarte Lopes, Sérgio Neves Monteiro and Carlos Mauricio Fontes Vieira

Abstract In the last years, the research on the natural materials has been growing, trying to find alternatives instead of synthetic materials. Among these materials are the lignocellulosic fibers or more specifically cotton fiber. This work evaluated the performance of composites with polyurethane as matrix, reinforced by cotton fabric when tested in impact at Izod configuration. The fabric volumes amount was from 10 up to 30%, aligned with the matrix. Good results at impact resistance were observed in specimens with 10%vol of cotton fabric. For 20%vol samples, a significant drop at impact energy absorption was caused by some irregularities between fabric and matrix.

Keywords Cotton fabric · Polyurethane matrix · Izod impact test

Introduction

Natural lignocellulosic fibers such as cotton, flax, jute, and ramie have long been used for fabric manufacture. However, the development of synthetic fibers brought new options for the textile industry and began to compete advantageously with natural fibers. In addition to the scientific and environmental aspects, there are also social aspects that contribute to use natural fiber composites, for example, the cultivation of these is a source of income generation in communities in the north and northeast of Brazil, among others.

C. G. D. Ribeiro · J. P. R. G. de Carvalho (✉) · F. P. D. Lopes · C. M. Fontes Vieira
Advanced Materials Laboratory – LA-MAV, State University of the Northern Rio de Janeiro – UENF, Avenida Alberto Lamego, 2000, Campos dos Goytacazes 28013-602, Brazil
e-mail: julianarufino.arq@gmail.com

S. N. Monteiro
Materials Science Department, Military Institute of Engineering – IME, Praça General Tibúrcio, 80, Urca, Rio de Janeiro, RJ 22290-270, Brazil

S. Ikhmayies et al. (eds.), *Green Materials Engineering*, The Minerals, Metals & Materials Series, https://doi.org/10.1007/978-3-030-10383-5_26

The advantages to use natural fibers over traditional reinforcing materials, such as glass fibers, talc and mica are: low cost; high tenacity; good mechanical and thermal properties; reduction of machine wear; easily of separation and biodegradability, among others. Wood fiber reinforced composites with specific properties similar to glass fiber polypropylene composites are already reported in the literature [1, 2].

Cotton is a white fiber that grows around the seeds of some species of the genus Gossypium of Malvaceae family, and its fibers are harvested manually or with the machine auxiliary. Being that manual, the collection form is usually done in the tree and brings a product without impurities. Being the most used natural fiber in fabrics, cotton is basically composed of cellulose, containing only 3–15% of non-cellulosic material. Cotton fibers represent about 90% of the natural fibers used in Brazil [3].

To obtain the cotton fabric, it is necessary to subject the fiber to various processes such as ginning that consists the separation between fiber and seed, done manually or with the machine harvesting, followed by spinning which is the step in which the cotton yarns are made from the fiber, with different thicknesses. Next, the weaving process is used, being one of the most important processes to obtain the fabric, also called loom, two yarns are used simultaneously to give weft and weave to the fabric. After the woven fiber, the cotton goes through several other phases until it becomes the final product. Subsequently, the scorching is the process of flame passing through the fabric to eliminate excess fibers to improve the visual appearance and also the touch. Thereafter, the process known as bleaching submits the fabrics into bleaches to become lighter and lighter, especially, if the natural fibers are yellowish or with a lot of variation. Mercerization is the cold application of caustic soda that reacts with cotton cellulose and increases the strength, brightness, durability, and flexibility of the fabric. And dyeing consists in changing the color of the tissue by means of dye treatment. Finally, the fabric goes through various chemicals to gain resistance and protection against harmful agents [4].

The vegetable polyurethane resin based on castor oil is a bicomponent, 100% solid- and solvent-free, not releasing toxic vapors. It is formulated by the cold blending of a prepolymer (component A) and a polyol (component B), resulting in polymers with different characteristics of exceptional physical–chemical stability. Castor oil besides being abundant in Brazil is a renewable raw material and does not cause aggression to the environment. Composites based on polyurethane reinforced with linen and jute fabric and obtained a combination of good engineering properties and low density [4].

In view of the above, the present work has the objective of evaluating the performance of castor oil-based vegetable polyurethane matrix composites reinforced with 10–30%vol cotton fabric for the absorption of Izod impact energy.

Materials and Methods

Determination of Cotton Fabric Density

Pycnometry is a laboratory technique developed to determine the density and specific mass of liquids using a pycnometer that consists of a container made of suitable material and having its volume accurately determined, shown in Fig. 1. As a reference density, the specific mass of water is used, because one liter of water weighs one kilogram at ambient pressure and at 25 °C, that is, 1 g/cm^3 [5].

The density calculation in relation to the distilled water was carried out using a pycnometer and a digital scale with an accuracy of ±0.001 g. First, the tare of the digital scale was carried out with an empty pycnometer, and then the mass of the pycnometer + water was weighed. After that, the mass of the pycnometer + solid + distilled water was measured, thus obtaining the apparent density of the tissue in relation to the water. The cotton cloth was cut into approximately 1 cm pieces to facilitate its insertion into the pycnometer.

Before placing the fabric in the pycnometer with distilled water, it was kept under vacuum for 2 h in order to remove the existing bubbles. Parallel to the test, the moisture correction factor that could exist in the material was calculated. For this,

Fig. 1 Pycnometer

Fig. 2 Fabric maintained in vacuum desiccator (left) and pycnometer filled with water and cotton cloth (right)

a quantity of fabric was weighed and placed in a stove for 24 h and soon after; the cooling was done under vacuum in order to avoid the absorption of the humidity of the medium, shown in Fig. 2.

The expression used to determine the density (*ds*) of the fabric as a function of water density was given by Eqs. 1 and 2, where m1 is the mass of the empty pycnometer; m2 is the mass of the pycnometer with the fiber; m3 is the mass of the pycnometer with the fiber + water, and m4 is the mass of the pycnometer with water.

$$ds = \frac{\rho\mathrm{s}}{\rho\,\mathrm{water}} = \frac{\mathrm{m2} - \mathrm{m1}}{(\mathrm{m4} - \mathrm{m1}) - (\mathrm{m3} - \mathrm{m2})} \quad (1)$$

$$\rho\mathrm{s} = ds \cdot \rho\,\mathrm{water} \quad (2)$$

As the scale was calibrated with the pycnometer, the calculation was performed considering that the mass of the empty pycnometer was zero. To raise the density, a 50 ml pycnometer was used.

Samples Preparation

The cotton fabrics were donated by a fabric-cutting factory, located in Campos dos Goytacazes/RJ-Brasil. The vegetable polyurethane resin was obtained from IMPERVEG®.

In the preparation of polymer matrix samples, the resin AGT 1315 that is based on vegetable polyurethane (originating from castor oil) was used. The density provided by the manufacturer was 1.05 g/cm^3 and the volumetric ratio between component A (prepolymer) and B (polyol) is 1: 1.2 by weight.

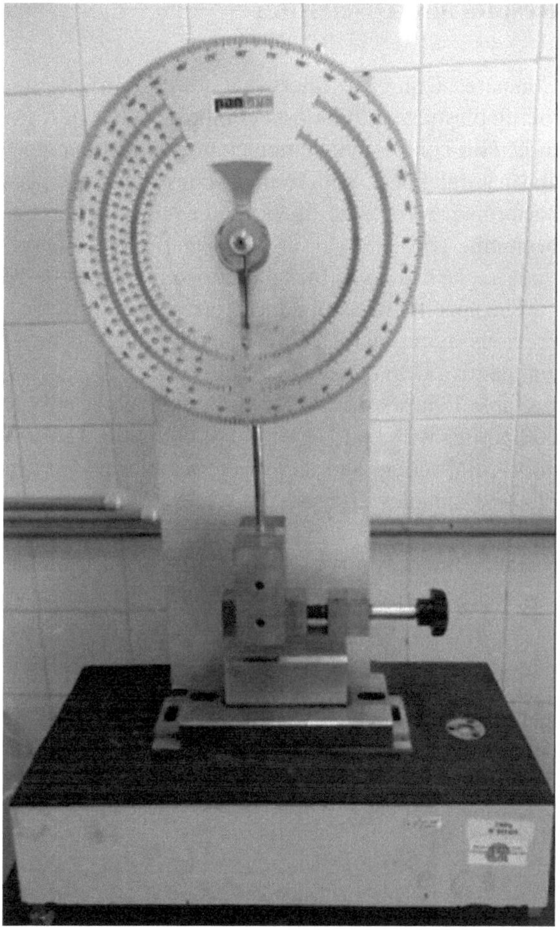

Fig. 3 Izod test analog pendulum

After mixing the resin in the proportions pre-defined by the supplier, it was poured onto a rectangular metal mold (15 cm × 13 cm × 1 cm), permeating the fabric strips (15 × 13 cm), in order to obtain plates of composites with 10, 20, and 30%vol of cotton fabric. After the matrix was filled, it was pressed to the press with 2 ton application, in order to improve the matrix/fabric interface, and accelerate the catalysis through an internal heating caused by the application of pressure. After 24 h the plates were removed from the press, demolded and taken to the oven at 100 °C for 48 h to perform the post-cure.

After the drying and post-cure time, the plates were cut in the dimensions of 63 mm × 12.7 mm × 10 mm to obtain the Izod Impact specimens, with a notch spacing of 22 mm, 2.54 mm depth, following the guidelines of ASTM D256. The izod impact test was performed using a Pantec Machine, Fig. 3.

Results and Discussion

Compared with glass fibers that are used in 95% of the cases for reinforcement in thermoplastic and thermoset composites, recent research shows that the application in certain composites demonstrates competitive performance between glass fibers and natural fibers, and further states that after decades of development of high-performance synthetic fibers such as carbon, glass, and aramid, natural fibers have been emerging and gaining more and more interest, especially when it comes to replace glass fibers in the automotive industries [6–8].

The specific mass obtained for the cotton fabric was 1.39 g/cm^3, which attests previous studies, which defend natural fibers as materials of low density and highly competitive with the synthetic fibers.

Table 1 shows the results of Izod impact energy for pure polyurethane resin and composites with up to 20%vol cotton fabric. The samples of 30%vol did not obtain good compaction, and there were bubbles and irregularities that prevented the test of these samples. Subsequently, new samples of this volumetric fraction will be prepared.

Based on the results of Table 1, the variation in impact energy for each volumetric fraction is shown in Fig. 4.

Table 1 Izod impact energy for composites with polyurethane resin matrix reinforced by cotton fabric

Fabric volume fraction (%)	Impact energy (J/M)
0	184,26
10	188,34
20	145,90

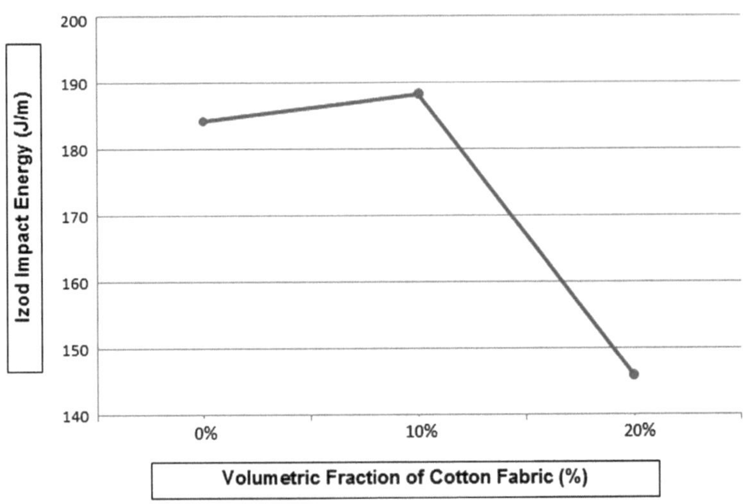

Fig. 4 Izod Impact energy as a function of volumetric fraction of cotton fabric

It is possible to notice that occurred a significant drop at Izod impact energy absorption for the 20%vol samples with cotton fabric, possibly caused by the difficulty of preparing the samples uniformly aligned and in order to reduce irregularities in the interface between the fiber and the polymer.

Conclusions

The density of the cotton fabric obtained by the pycnometry method was 1.39 g/cm^3.

Impact tests with polymer matrix composites reinforced with 10%vol of cotton fabrics perform well in the absorption of impact energy compared to samples of pure polyurethane resin.

As for the impact test, it is observed that the addition of cotton fabric increases the impact resistance of the composites. However, this effect is reduced when good compaction is not achieved between the fabric/matrix interface due to irregularities that reduce the absorption of impact energy due to the formation of bubbles inside the samples.

Acknowledgements The authors are grateful for the support of the Laboratory of Advanced Materials—LAMAV UENF, mainly to Renan and soils laboratory (LSOL), located in CCTA/UENF.

References

1. Krishnan M, Narayan R (1992) Compatibilization of biomass fiber with hydrophobic materials. In: Materials research proceedings, p 266
2. Rowell RM, Schultz TP, Narayan R (1992) Emerging technologies for materials and chemical from biomass. In: ACS symposium series, p 476
3. Relatório de Diagnóstico do Ministério do Desenvolvimento (2002) Indústria e Comércio (MDIC). Senai-Cetiqt, Programa Brasileiro de Prospectiva Tecnológica Industrial
4. Chierice GC, Neto SC, Sabatini A (1995) Ricinoquimica–Óleo de Rícino Fonte renovável Vegetal e seus Derivados para o Uso Industrial na Aplicação Automotiva. Instituto de Química de São Carlos, USP
5. Montiny (2012) In: Picnometria. https://ipemsp.wordpress.com/2012/04/02/picnometria. Accessed 30 Aug 2018
6. Bledzki AK, Zhang W, Chate A (2001) Natural-fiber-reinforced polyurethane microfoams. Compos Sci Technol 2405
7. Mohanty AK, Misra M, Drzal LT (2005) Natural fibers, biopolymers, and biocomposites. [s.l.]. CRC Press
8. Saheb DN, Jog JP (1999) Natural fiber polymer composites: a review. Adv Polym Technol 18(4):351–363

Performance of Natural Curaua Non-woven Fabric Composites as Stand-Alone Targets Against Standard 9 mm and 7.62 mm Projectiles

Fabio de Oliveira Braga, Michelle Souza Oliveira, Fábio da Costa Garcia Filho, Sergio Neves Monteiro and Édio Pereira Lima, Jr

Abstract Natural fiber (NF) reinforced composites are gaining increasing attention from researchers and engineers, due to properties such as high strength, high toughness, low density, and biodegradability. More recently, their ballistic properties are being studied, aiming the application as low-weight/low-cost armor shields, either as stand-alone plates or integrating multilayered armor systems. Among the NF, stand out the curaua, the fibers extracted from the leaves of the *Ananas Erectifolius*, a plant from Amazon region. Their strength and Young's modulus are among the highest of the NF, almost comparable to the glass fibers. In the present work, curaua non-woven fabric composites (CNWFC) were tested as stand-alone targets against 9 mm and 7.62 commercial ammunition bullets. The residual velocity (V_r) of the projectiles after impacting the targets was used as measure of the ballistic performance. A greater structural damage was observed in the composites tested with 9 mm, although these bullets have a much lower Kinect energy. Therefore, a higher energy absorption was observed, and thus, the protection against 9 mm projectiles might be more promising application for the CNWFC stand-alone plates.

Keywords Composite · Natural fiber · Curaua fabric · Ballistic test

F. de Oliveira Braga
National Service of Industrial Apprenticeship – SENAI, Rio de Janeiro, Brazil

F. de Oliveira Braga (✉)
Fluminense Federal University - UFF, Niterói, Rio de Janeiro, Brazil
e-mail: fabio_obraga@yahoo.com.br

M. S. Oliveira · F. da C. Garcia Filho · S. N. Monteiro · É. P. Lima, Jr
Military Institute of Engineering - IME, Rio de Janeiro, Brazil

S. Ikhmayies et al. (eds.), *Green Materials Engineering*, The Minerals, Metals & Materials Series, https://doi.org/10.1007/978-3-030-10383-5_27

Introduction

Natural fiber (NF) reinforced composites are gaining increasing attention from researchers and engineers, due to properties such as high strength, high toughness, low density, and biodegradability [1–3]. More recently, their ballistic properties are being studied, aiming the application of these materials as low-weight/low-cost armor shields, either as stand-alone plates or integrating multilayered armor systems [4–7].

Among the NF, stand out the curaua, the fibers extracted from the leaves of the *Ananas Erectifolius*, a plant from the Amazon region. Their strength and Young's modulus are among the highest of the NF, almost comparable to the glass fibers [8].

Additionally, the ballistic properties of the curaua composites have been investigated, and the results were considered promising [9–12]. The first investigation was conducted by Monteiro et al. [9], who studied epoxy matrix composites reinforced with continuous and aligned curaua fibers, as part of multilayered armor systems (MAS). After that, the same group [10] studied similar curaua fiber reinforced composites (CuFRC) as part of a MAS, this time using an unsaturated polyester resin as polymeric matrix and varying the fiber fraction from 0 to 30 vol. %. They found out that the increasing of fiber fraction improves the integrity of the specimen after the ballistic impact, which is directly related to the higher impact strength of the composite [10]. Braga et al., on the other hand, [11] studied the behavior of CuFRC as stand-alone ballistic targets. Also, Braga et al. [12] performed MAS ballistic tests using curaua non-woven fabric composites (CNWFC) as intermediate layer in a MAS similar to that investigated by Monteiro et al. [9]. The results were quite promising, and so it might be important to investigate those composites in other ballistic applications.

Therefore, the objective of the present work is to study the behavior of CNWFC as stand-alone ballistic armors against either 7.62 or 9 mm ammunition.

Materials and Methods

The curaua non-woven fabric was acquired from the Brazilian company Permatec Triangel, as large mats with areal density of approximately 0.830 kg/m^2. The mats were cut in small pieces and dried at 60 °C for 24 h for the production of composites. The target plates were then prepared by compression molding at room temperature (25 °C), in the volumetric fraction of 30%. Two different thickness plates were prepared, 10 and 16.5 mm. The fiber fabrics were carefully positioned in the mold, in layers, intercalating with the resin–hardener mixture. The mixture was left resting under the pressure of 3 MPa for 24 h to make sure that the curing reaction ended. The polymeric matrix was an epoxy diglycidyl ether of bisphenol A (DGEBA) resin, hardened with triethylene tetramine (TETA, 13 phr), both produced by Dow Chemical and acquired from Brazilian company Epoxyfiber.

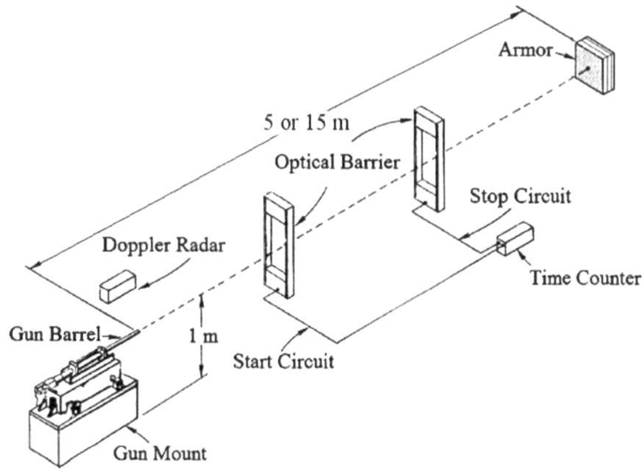

Fig. 1 Experimental arrangement of the ballistic tests

The targets were subjected to ballistic impact at the Army Assessment Center (CAEx), in the Marambaia Peninsula, Rio de Janeiro. The shooting device was a model B290 HPI (High Pressure Instrumentation), which consists in a gun barrel with laser sight. For the velocity measurements, it has been employed a model SL-520P Weibel Doppler radar provided with a Windopp software to process the radar raw data. The ammunition was either 7.62 mm M1 (9.7 g) or 9 mm (8 g), both commercial. The targets were firmly positioned either 15 m away from the gun barrel, for the 7.62 mm shootings, or 5 m away, for the 9 mm. All the shootings were performed with the bullet following a trajectory perpendicular to the target. A schematic diagram of the experimental apparatus is shown in Fig. 1.

The projectile's velocity was measured immediately before (V_i) and after (V_r) the impact. The kinetic energy variation of the projectile was related to the energy absorbed by the target (E_{abs}) and used for comparison between the materials. It can be calculated by Eq. 1.

$$E_{abs} = \frac{m(V_i - V_r)^2}{2} \tag{1}$$

where: m = mass of the projectile.

Two other calculations were performed: The percentage of the projectiles energy absorbed by the target ($\%E_{abs}$), Eq. 2, and the limit velocity (V_L), which is an estimation of the velocity that the target would be able to stop the projectile, since V_r is considered to be zero.

$$\%E_{abs} = \frac{E_{abs}.100\%}{E_i} \tag{2}$$

$$V_L = \sqrt{\frac{2.E_{abs}}{m}} \tag{3}$$

where: $E_i = \frac{mV_i^2}{2}$ = Impact energy.

Results and Discussion

Figure 2 shows a fitted curve obtained by treating the raw Doppler data with the software Windopp. This particular result represents the drop in the 7.62 mm projectile's velocity with time, before and after impacting the 10 mm thick target. In Fig 2, the point of impact is around 0.02 s. Similar curves were obtained for all the shootings. Based on these graphs, it was possible to obtain the parameters V_i and V_r, which are presented in the Table 1 for the 10 mm thick plate subjected to 7.62 mm impact. The E_{abs} (Eq. 1), % E_{abs} (Eq. 2), and V_L (Eq. 3) calculations are also presented in Table 1.

Table 1 shows that a very small amount of the projectile's energy (V_i) is absorbed by the target. This is attributed to the high Kinect energy of the 7.62 mm projectile, due to its velocity of about 850 m/s. Figure 3 shows the aspect of the specimens in the region around the perforation.

Table 2 shows the ballistic test parameters for the 16.5 mm thick composite plates subjected to 7.62 mm ballistic impact.

As expected, the thicker 16.5 mm plate was able to absorb a greater amount of the projectile's energy (~6.8%, Table 2) than the 10 mm plate (~4.0%, Table 1). The difference in E_{abs} can be considered small, and thus, it is expected the same fracture aspect around the perforations. Figure 4 confirms this similarity.

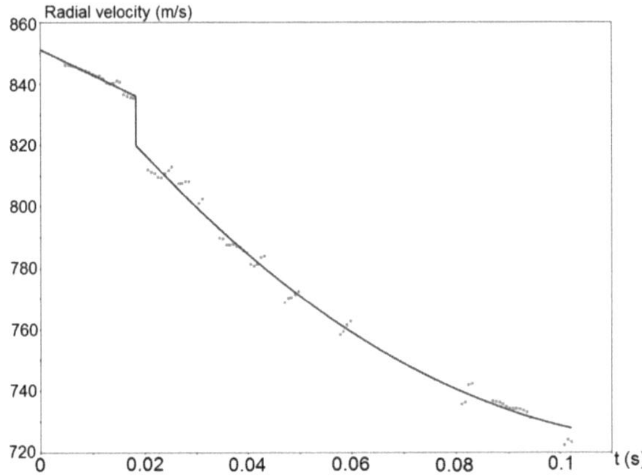

Fig. 2 Variation of projectile velocity with time in the shooting event

Table 1 Measured parameters of the 10 mm thick CNWFC ballistic test using 7.62 mm ammunition

	V_i (m/s)	V_r (m/s)	E_{abs} (J)	$\%E_{abs}$ (%)	V_L (m/s)
	836.030	819.899	129.552	3.822	163.437
	836.777	821.841	120.150	3.538	157.395
	840.802	823.650	138.461	4.038	168.964
	836.366	816.053	162.793	4.798	183.209
Média	837.5 ± 2	820.4 ± 3	137.7 ± 18.3	4.0 ± 0.5	168.2 ± 11.0

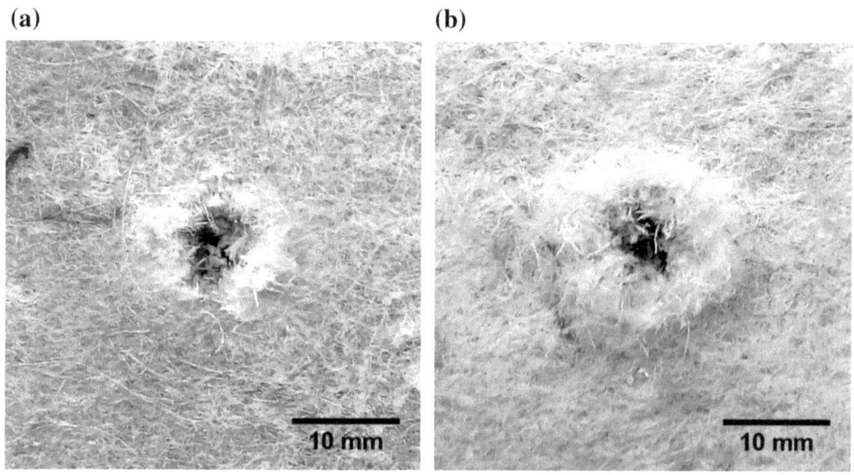

Fig. 3 Aspect of the perforations in the 10 mm thick plate of CNWFC subjected to 7.62 mm ballistic impact: **a** front side; **b** rear side

Table 2 Measured parameters of the 16.5 mm thick CNWFC ballistic test using 7.62 mm ammunition

	V_i (m/s)	V_r (m/s)	E_{abs} (J)	$\%E_{abs}$ (%)	V_L (m/s)
	855.943	827.970	228.455	6.429	217.035
	855.960	824.800	254.007	7.148	228.850
	847.523	818.776	232.320	6.669	218.863
	842.572	811.265	251.117	7.293	227.545
Média	850.5 ± 6.6	820.7 ± 7.4	241.5 ± 12.9	6.8 ± 0.4	223.1 ± 5.9

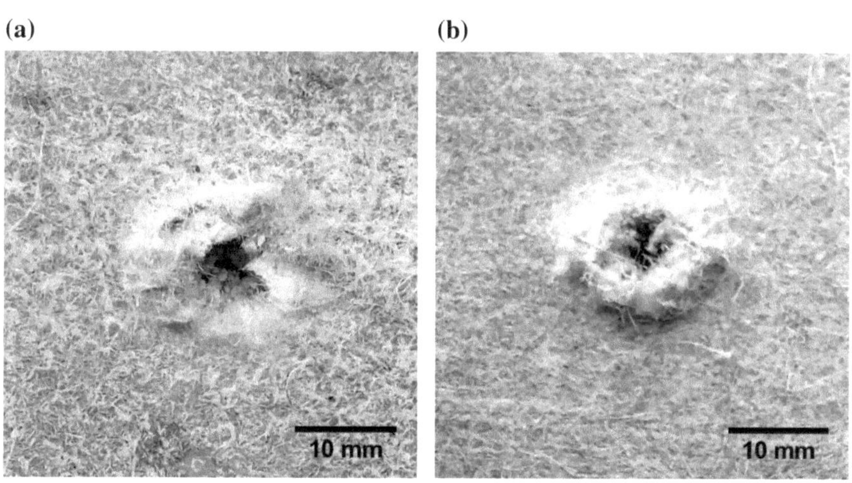

Fig. 4 Aspect of the perforations in the 16.5 mm thick plate of CNWFC subjected to 7.62 mm ballistic impact: **a** front side; **b** rear side

Table 3 Measured parameters of the 16.5 mm thick CNWFC ballistic test using 9 mm ammunition

	V_i (m/s)	V_r (m/s)	E_{abs} (J)	$\%E_{abs}$ (%)	V_L (m/s)
	334.191	245.708	205.245	45.943	226.520
	334.321	242.930	211.022	47.200	229.686
	332.955	234.794	222.923	50.272	236.074
	337.715	239.147	227.441	49.855	238.454
Média	334.8 ± 2.0	240.6 ± 4.7	216.6 ± 10.2	48.3 ± 2.1	232.7 ± 5.5

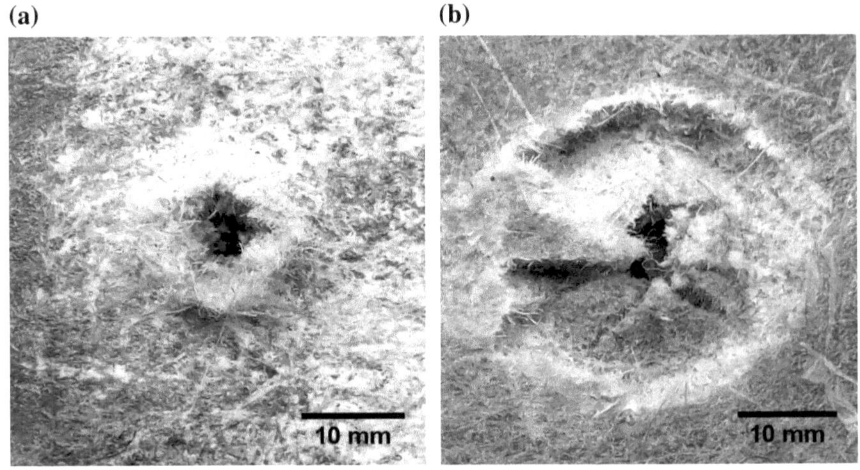

Fig. 5 Aspect of the perforations in the 16.5 mm thick plate of CNWFC subjected to 9 mm ballistic impact: **a** front side; **b** rear side

Table 3 shows the ballistic test parameters for the 16.5 mm thick composite plates subjected to 9 mm ballistic impact. In this case, the projectile's V_i is much smaller, so the target is able to absorb a significant amount of the energy (~48%, Table 3). The fracture aspect is also significantly different, as can be seen in Fig. 5.

The fracture aspect of the rear side of the composite indicates that a larger region of the plate responded to the impact, explaining the greater energy absorption. The fracture mechanism might include the tensile failure of the fibers and shear (plug).

Summary and Conclusions

- The behavior of curaua non-woven fabric composites (CNWFC) as stand-alone ballistic armors has been studied either 7.62 or 9 mm ammunition.
- A very small amount of the 7.62 mm projectile's energy could be absorbed by the 10 and 16.5 mm thick CNWFC targets (4.0 ± 0.5 and 6.8 ± 0.4%, respectively).

This is attributed to the high Kinect energy of the 7.62 mm projectile, due to its velocity of about 850 m/s.
- A significantly higher amount of the projectile's energy could be absorbed from the 9 mm projectile by the 16.5 mm thick CNWFC target (48.3 ± 2.1%). This is attributed to a change in the mechanisms of fracture. In this case, larger region of the plate responded to the impact.

Acknowledgements The authors thank the Brazilian agencies CAPES and CNPq for the financial support and CAEX for performing the ballistic tests.

References

1. Güven O, Monteiro SN, Moura EAB, Drelich JW (2016) Re-Emerging field of lignocellulosic fiber–polymer composites and ionizing radiation technology in their formulation. Polym Rev 56(4):702–736
2. Pickering KL, Efendy MGA, Le TM (2016) A review of recent developments in natural fibre composites and their mechanical performance. Compos Part A 83:98–112
3. Chandrasekar M, Ishak MR, Sapuan SM, Leman Z, Jawaid M (2017) A review on the characterisation of natural fibres and their composites after alkali treatment and water absorption. Macromo. Eng 46(3):119–136
4. Benzait Z, Trabzon L (2018) A review of recent research on materials used in polymer—matrix composites for body armor application. J Compos Mater. https://doi.org/10.1177/0021998318764002
5. Monteiro SN, Milanezi TL, Louro LHL, Lima EP Jr, Braga FO, Gomes AV, Drelich JW (2016) Novel ballistic ramie fabric composite competing with Kevlar™ fabric in multilayered armor. Mater Design 96:263–269
6. Monteiro SN, Candido VS, Braga FO, Bolzan LT, Weber RP, Drelich JW (2016) Sugarcane bagasse waste in composites for multilayered armor. Eur Polym J 78:173–185
7. Akubue PC, Igbokwe PK, Nwabanne JT (2015) Production of Kenaf fibre reinforced polyethylene composite for ballistic protection. IJSER 6(8):1–7
8. Oliveira FH, Helfer AL, Amico SC (2012) Mechanical behavior of unidirectional curaua fiber and glass fiber composites. Macromol Symp 319:83–92
9. Monteiro SN, Louro LHL, Trindade W, Elias CN, Ferreira CL, Lima ES, Weber RP, Suarez JCM, Figueiredo ABS, Pinheiro WA, Silva LC, Lima EP (2015) Natural curaua fiber-reinforced composites in multilayered ballistic armor. Metall Mater Trans A 46(10):4567–4577
10. Monteiro SN, Braga FO, Lima EP, Louro LHL, Drelich JW (2016) Promising curaua fiber-reinforced polyester composite for high-impact ballistic multilayered armor. Polym Eng Sci 57(9):947–954
11. Braga FO, Bolzan LT, Lima Jr. EP, Monteiro SN (2017) Performance of natural curaua fiber-reinforced polyester composites under 7.62 mm bullet impact as a stand-alone ballistic armor. J Mater Res Technol 6(4):323–328
12. Braga FO, Bolzan LT, Luz FS, Lopes PHLM, Lima EP, Monteiro SN (2017) High energy ballistic and fracture comparison between multilayered armor systems using non-woven curaua fabric composites and aramid laminates. J Mater Res Technol 6(4):417–422

Reuse of Quarry Waste in Artificial Stone Production with Using Vacuum, Compression, and Vibration

Elaine A. S. Carvalho, Juan P. B. Magalhães, Rubén J. S. Rodriguez, Eduardo A. Carvalho, Sergio N. Monteiro and Carlos Mauricio Fontes Vieira

Abstract Due to the natural resource depletion, the incorporation of quarry dust, a by-product from the stone milling and abundantly available in the artificial stone production, can represent an ecologically viable method to reduce environmental problems besides adding it a value. In this work, it was used 90% of stone dust mass and 10% of epoxy resin mass for producing a new artificial ornamental stone using vacuum, vibration, and compression which allow its use on civil construction. For preparing the specimens, the granulometric composition associated to the best waste packaging has been determined by the simplex method. The characterization consisted on analyzing its physical and mechanical properties. The artificial stone developed presented mechanical properties according to the expected pattern, with break tension of 32 MPa, and when evaluated the break tension by flexion, the specimens passed through freezing and thaw process without any change, being able to be used in environments with low temperature. The results found for density, water absorption, and porosity were according to the patterns for civil construction, having porosity lower than 0.5%.

Keywords Artificial stone · Residue · Quarry dust · Mechanical properties

Introduction

The industrial waste elimination is a global problem. Besides, its non-biodegradable nature increases the problem, and the people's quality of life depends now on the use of alternative products that minimize the environmental impact [1].

E. A. S. Carvalho (✉) · J. P. B. Magalhães · R. J. S. Rodriguez · E. A. Carvalho · C. M. Fontes Vieira
Advanced Materials Laboratory – LAMAV, State University of the Northern Rio de Janeiro – UENF, Av. Alberto Lamego, 2000, Campos dos Goytacazes 28013-602, Brazil
e-mail: elainesanttos@yahoo.com.br

S. N. Monteiro
Military Institute of Engineering – IME, Materials Science Department, Praça General Tibúrcio 80, Urca, Rio de Janeiro, RJ 22290-270, Brazil

S. Ikhmayies et al. (eds.), *Green Materials Engineering*, The Minerals, Metals & Materials Series, https://doi.org/10.1007/978-3-030-10383-5_28

The quarry industrial practice generates powder particles during the operations of milling, cutting, and sieving when producing the stones for civil construction. This technique, which represents around 10% of the processed stone and gravel blocks, is generally discarded in outdoor, rivers, and lagoons piles, which is not an environmentally correct solution [2, 3]. To minimize the impacts on the nature, it has been studied the development of materials, known as an artificial stone, produced by stone wastes and polymeric resin [4].

The natural stones present a series of limitations for its use, such as high cost, susceptibility to blemishes, and ease of breaking. The synthetic stone, according to the most highlighted companies that develop the artificial stone and construct equipments for its manufacture, is a material with good mechanical performance besides being waterproof and non-staining. It happened due to the liquid penetration difficulties, keeping it only on the surface as the resin not only makes adherence between the stone particles but it also penetrates between its pores [5].

Despite the fact that Brazil is one of the world's largest natural ornamental stone producers, the country has not been yet highlighted when it comes to artificial stone. A reflection about it consists of the large volume of artificial stones imported annually. Related to economic values, according to Abirochas 12/2014, imports of artificial stones remained near to those of natural materials and, unlike them, there was a significant increase in physical volume (+31.42%). Artificial stones showed imports of USD 43.9 million and 48.1 thousand tons in September 2014. Its average price (USD 890.8/t) also remains higher than that of natural materials also imported (USD 705 8/t) [6].

Researchers' works are being dedicated on scientific solutions and techniques for the residue reuse. The main idea is to remove it from the environment with low cost and, if possible, in a profitable way. A typical solution would be the addition of these products to a polymeric matrix. This has become especially profitable for those which simulate natural ornamental stones and are commercialized artificial stones [7, 9].

The main objective is to produce a new product using vacuum, vibration, and compression techniques, which contributes not only to reduce the amount of industrial wastes produced daily, but also to combine technical and economical advantages for applications in the civil construction.

Materials and Methods

The residue was collected from a pebble separation process in the Itereré quarry located in the mountain region of Serra da Bela Vista, located 17 km from the city of Campos dos Goytacazes, north of Rio de Janeiro state, Brazil. The polymer used as matrix was the diglycidyl ether of the bisphenol A (DGEBA) epoxy resin mixed with stoichiometric phr = 13 triethylene tetramine (TETA) as hardener, both from Epoxyfiber, Brazil. The residues were sieved according to ABNT NBR 7181 [10] and classified in three granulometric ranges: large (from 2 mm up to 0.42 mm),

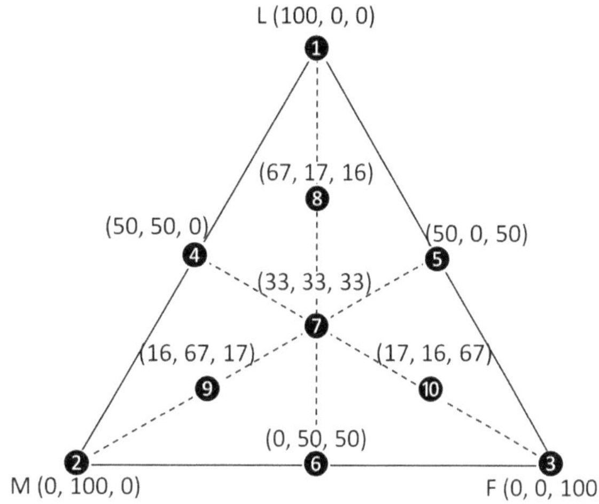

Fig. 1 Ternary diagram of the simplex complete cubic model reproduced with permission from (Carvalho et al. [9])

medium (from 0.42 mm up to 0.075 mm), and fine particles (with grains with size inferior to 0.075 mm). Based on these three granulometric classifications, 10 distinct mixtures were proposed for each type of residue for best-packed condition using the simplex-lattice design (SLD) numerical modeling methodology [9]. Figure 1 shows schematically the mixture compositions investigated.

The determination of best-packed composition for the quarry dust was associated with the highest dry apparent density. This density was obtained according to the Brazilian NBR 3388 norm [11] for 10 different compositions considered in the simplex method, Fig. 1. For each composition, three samples were used to assure statistical validation. Each sample of quarry dust compositional mixture was placed in a steel vessel and allowed to vibrate for 2 min under a load of 10 kg. The mixture was weighed, and the apparent density was calculated.

A minimum resin rate was calculated, used to fill the porosity within the particles, and thus, the mixture utilized was 90% particles and 10% epoxy resin. The artificial stone quarry dust (ASQ) plates were produced with dimensions 100 mm × 100 mm × 10 mm by the vacuum, vibration, and compression method. Initially, the residues were stove dried at 100 °C for 24 h to release moisture. The mixture was put in vacuum 600 mmHg; then, the mold was vibrated for 2 min and a compression pressure 10 MPa, with a curing temperature of 90 °C for 20 min. Later on, the plates were cut in the dimensions specified in the tests performed to characterize it.

The density, water absorption, and apparent novel stone artificial porosity were determined as per Brazilian NBR 15.845 norm [12]. Ten prismatic specimens, cut from the plate with dimensions of 100 × 25 × 10 mm, were three-point bend tested in a model 5582 Instron machine following the recommendation for agglomerated stones as per the Spanish EN 14617 standard [13] as well as the Annex F of the Brazilian NBR 15.845 standard [12].

An important feature when the stone is destined for exportation to countries with cold weather is its resistance fall related to the freezing cycles and thaw. To evaluate

Table 1 Weight and density of the quarry waste particle mixtures

Mixture number in Fig. 1	Weight (g)	Density (g/cm^3)
1	1.447 ± 0.013	1.435 ± 0.010
2	1.703 ± 0.028	1.688 ± 0.030
3	1.490 ± 0.009	1.476 ± 0.010
4	1.635 ± 0.015	1.613 ± 0.015
5	1.817 ± 0.043	1.793 ± 0.045
6	1.813 ± 0.003	1.790 ± 0.010
7	1.832 ± 0.041	1.810 ± 0.044
8	1.867 ± 0.033	1.843 ± 0.031
9	1.475 ± 0.043	1.457 ± 0.040
10	1.793 ± 0.013	1.770 ± 0.010

the effect of the cycles, it was followed orientation of D annex of the Brazilian norm NBR 15.845 (2010) [12] It was evaluated the alteration in the resistance by flexion in three slabs of artificial stone (ASQ) after the freezing and thaw cycles.

Results and Discussion

Table 1 presents the weight and density of the dried particle mixtures indicated in Fig. 1. In this table, it highlights the mixture number 8, which displays both the greatest weight and density. Consequently, this mixture also has the close-packing highest granite particles. The mixture number 8, with 67% large, 16% medium, and 16% fine particles, was selected as most closely packed in spite of different geometric shapes [14].

Table 2 presents the experimental results found for: density, water absorption, and porosity of the new artificial stone developed with 90% of resin and vacuum use. The apparent density value of 2.33 g/cm^3 is under of the artificial stones used for coating commercialized between 2.4 and 2.5 g/cm^3 [7], what means a lighter material for transportation. About the water absorption, industrial artificial marble is produced from 0.09 to 0.40% [9]. The stone developed in this present work obtained 0.14% comparing to the artificial marbles produced having a result lower than the one related by Carvalho et al. [9], that used the same material with 10 and 15% resin epoxy, but without the vacuum use, obtaining 0.35 and 0.25% of water absorption [9]. Borselino et al. [15] used epoxy resin and marble wastes, and the water absorption value was of 0.25% [15]. Ribeiro et al. [8] also used marble waste and polyester resin and obtained the average of 0.19% [8].

Chiodi and Rodriguez [16] classified the materials for construction and determined its high quality when presenting porosity values under 0.5% [16]. The porosity found for the artificial stone produced by gravel waste and epoxy resin was 0.32% determining a high-quality product. It justifies that the use of vacuum during the production

Table 2 Physical properties of apparent density, water absorption, and apparent porosity as well as mechanical properties of the investigated artificial stones

Property	Artificial stone quarry dust produced (ASQ)
Apparent density (g/cm^3)	2.33 ± 0.02
Water absorption (%)	0.14 ± 0.02
Apparent porosity (%)	0.32 ± 0.04
Flexural strength (MPa)	32.00 ± 1.98
Flexural strength (MPa)/with freeze and thaw cycles	31.00 ± 1.10

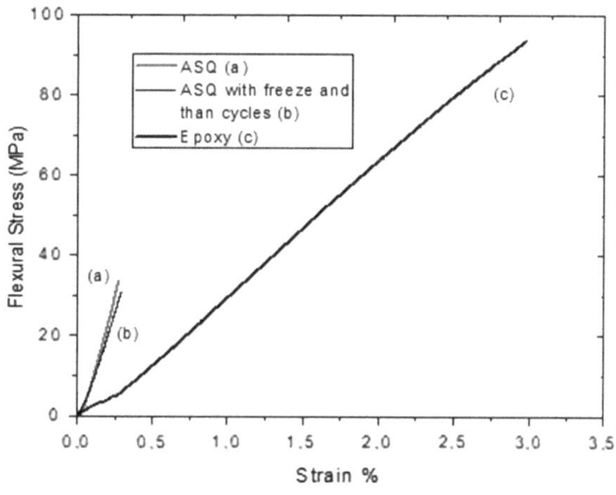

Fig. 2 Flexural stress x strain curves for neat epoxy, ASQ, and ASQ with freeze and thaw cycles reproduced with permission, the value of the neat epoxy [17]

process decreases the porosity levels [4], proving by the work of Carvalho et al. [9] that used 15% of resin without vacuum, and the value of porosity was of 0.55% [9].

Figure 2 shows the tension curve versus the pure epoxy resin deformation, reproduced with permission [17] with the new artificial stone (ASQ) with 10% of resin and the same with the cycle process of freezing and thaw. Comparing the curves, it is possible to see that the addition of charge caused a considerable increase in the material hardness, an expected behavior once the addition of hard particles by polymers usually acts increasing the elastic module [18].

It must be noticed that the maximum pure epoxy tension, 94 ± 5 MPa, is classified as higher than the new artificial stone developed 32 ± 1.98 MPa and in which suffered the freezing and thaw cycle 31 ± 1.10 MPa, an expected behavior as a reduction of resistance to the flexion is observed when huge dust quantities are added to the polymers [5].

Carvalho et al. [9] used the same material with 10 and 15% epoxy resin variation but without vacuum use, obtaining a maximum flexion tension of 30 ± 1 and 32 ± 3 MPa. The artificial stone developed with vacuum use with 5% less resin obtained the same result that Carvalho et al. [9] did with the use of 5% more resin [9]. Ribeiro et al. [8] obtained a value of 21.5 ± 1.9 MPa for the artificial stone produced with marble wastes, polyester resin, and solvent, probably because a solvent was used that decreases the chemical connections for the final product [8]. It is also important to mention that the ornamental stones used for civil construction are considered high-resistance materials when their tension and rupture exceed 20 MPa, like the new ASQ [16].

During the freezing and thaw cycles, no changes, fissures, or the detachment of parts on the specimens were observed. From the data presented (Table 2), it is observed that the values obtained for tension in flexion breaking of the specimens submitted to the freezing and thaw cycles are statistically the same obtained for the specimens which were not submitted to these cycles, indicating that the materials evaluated can be used in environments with temperature which cause the freezing to the water infiltrated into their pores, without suffering mechanical resistance drop.

Conclusions

- The quarry dust particle mixtures were determined based on the highest density, in terms of the best-packed characterization, found by means of the simplex-lattice design method.
- The values of density, 2.33 g/cm^3, water absorption, 0.14%, apparent porosity, 0.32%, and mechanical strength of 32 MPa for the ASQ are within the Brazilian standards and considered adequate when compared to other ornamentals stones used in civil construction.
- During the freezing and thaw cycles, it did not observe any change in the plates of artificial stone. It can be used in environments that promote freezing.
- The artificial stone produced demonstrated that the incorporation of quarry dust in its production for the civil construction can be an alternative and initiative for adding to the material value.

Acknowledgements The authors thank the support of this investigation by the Brazilian agencies: CNPq, CAPES, and FAPERJ. To technician Renan da Silva Guimarães.

References

1. Rana A, Kalla P, Verma HK, Mohnot JK (2016) Recycling of dimensional stone waste in concrete: a review. J Clean Prod. https://doi.org/10.1016/j.jclepro.2016.06.126
2. Lim SK, Tan CS, Li B, Ling T-C, Hossain M, Poon CS (2017) Utilizing high volumes quarry wastes in the production of light weight foamed concrete. Constr Build Mater 151:441–448
3. Ramos T, Matos AM, Schmidt B, João Rio J, Coutinho JS (2013) Granitic quarry sludge waste in mortar: effect on strength and durability. Constr Build Mater 47:1001–1009
4. Demartini TJC, Rodríguez RJS, Silva FS (2018) Physical and mechanical evaluation of artificial marble produced with dolomitic marble residue processed by diamond-plated bladed gang-saws. J Mater Res Technol. https://doi.org/10.1016/j.jmrt.2018.02.001

5. Lam dos Santos JP, Rosa LG, Amaral PM (2011) Temperature effects on mechanical behaviour of engineered stones. Constr Build Mater 25(1):171–174
6. Silva FS, Ribeiro CEG, Rodriguez RJS (2018) Physical and mechanical characterization of artificial stone with marble calcite waste and epoxy resin. Mater Res. http://dx.doi.org/10.1590/1980-5373-MR-2016-0377
7. Ribeiro CEG, Rodriguez RJS (2015) Influence of compaction pressure and particle content on thermal and mechanical behavior of artificial marbles with marble waste and unsaturated polyester. Mater Res 18:283–290
8. Ribeiro CEG, Rodriguez RJS, Vieira CMF, Carvalho EAS, Candido VS, Monteiro SN (2014) Production of synthetic ornamental marble as a marble waste added polyester composite. Mater Sci Forum 445–776:341–345
9. Carvalho EAS, Vilela NF, Monteiro SN, Vieira, CMF, Silva LC (2018) Novel artificial ornamental stone developed with quarry waste in epoxy composite. Mater Res. http://dx.doi.org/10.1590/1980-5373-MR-2017-1104
10. Brazilian Association of Technical Norms—ABNT (2016) ABNT NBR 7181: Soil – grain size analysis (in Portuguese). ABNT, Rio de Janeiro
11. Brazilian Association of Technical Norms—ABNT (1991) ABNT NBR MB 3388: soil – determination of minimum index void ratio of cohesionless soils – method of test (in Portuguese). ABNT, Rio de Janeiro
12. Brazilian Association of Technical Norms—ABNT (2010) ABNT NBR 15845-2: rocks for cladding—test methods (in Portuguese). ABNT, Rio de Janeiro
13. Spanish Association of Standards and Certification (2008) UNE-EN 14617-2-08: Test methods part 2: determination of the flexural strength (in Spanish). UNE-EN, Madrid
14. Alves HJ, Zauberas RT, Boschi AO (2010) Influence of granulometric distribution of clays on the dry milling yield in a hammer mill. Cerâmica 56:66–70 (in Portuguese)
15. Borselino C, Calbrese L, di Bella G (2009) Effects of powder concentration and type of resin on the performance of marble composite structures. Constr Build Mater 23:1915–1921
16. Chiodi FC, Rodriguez EP (2009) Application guide for rocks incoating. Abirochas, SP, Brazil [in Portuguese]
17. Carvalho EAS, Marques VR, Rodriguez RJS, Ribeiro CEG, Monteiro SN, Vieira CMF (2015) Development of epoxy matrix artificial stone incorporated with sintering residue from steelmaking industry. Mater Res 18:235–248
18. Lee M-Y, Ko C-H, Chang F-C, Lo S-L, Lin J-D, Shan M-Y et al (2008) Artificial stone slab production using waste glass, stone fragments and vacuum vibratory compaction. Cement Concr Compos 30:583–587

Reuse of the Iron Ore Residue Through the Production of Coating

Larissa Ribeiro, Elaine Carvalho, Maria Luiza Gomes, Mônica Borlini, Sergio N. Monteiro and Carlos Mauricio Fontes Vieira

Abstract In this work an artificial stone, which is a composite material, was produced using a reinforcement residue of iron ore and epoxy resin as matrix. This production can be a viable alternative to provide an alternative destination to the iron ore residue, since it has today dams and deposition ponds as destination, bringing risks to environment and to human health. The objective of this project was to produce artificial stone for coating, reusing iron ore residue, thus adding economic value to this material. The specimens were produced by pressing, with a resin content of 15%. Environmental and thermogravimetric analysis, physical tests, and chemical resistance experiments were performed. The residue was classified as non-inert and non-hazardous. The hydrochloric acid was the one that most attacked the artificial stone of iron ore residue (ASIO), with mass loss of 0.025 g. Regarding the water absorption, the ASIO obtained a value of 0.21%, value that is compatible with the ones found for artificial marbles. The production of artificial stone proved to be feasible, in addition to providing an environmentally favorable destination for the iron ore residue.

Keywords Artificial stone · Residue · Iron ore and coating

L. Ribeiro · E. Carvalho (✉) · M. L. Gomes · C. M. Fontes Vieira
Advanced Materials Laboratory—LAMAV, State University of the Northern Rio de Janeiro—UENF, Av. Alberto Lamego, 2000, Campos dos Goytacazes 28013-602, Brazil
e-mail: elainesanttos@yahoo.com.br

M. Borlini
Mineral Technology Center—CETEM, Rod. Cachoeiro, Km 05, Cachoeiro do Itapemirim, ES CEP: 29311970, Brazil

S. N. Monteiro
Materials Science Department, Military Institute of Engineering—IME, Praça General Tibúrcio, 80, Urca, Rio de Janeiro, RJ 22290-270, Brazil

S. Ikhmayies et al. (eds.), *Green Materials Engineering*, The Minerals, Metals & Materials Series, https://doi.org/10.1007/978-3-030-10383-5_29

Introduction

The stone market has been growing all over the world and Brazil has been contributing to meet the demand for ornamental stones, exporting it to 120 countries in 2016 mainly to USA, China, and Italy. In addition, Brazil imports artificial materials from 23 countries, with 82.6% coming from China [1, 2]. Thus, the market for creation of new materials in the area here in Brazil is very promising.

An artificial stone is a composite material, named commercially as industrialized stone, formed by a polymer matrix with 95% of natural aggregates incorporated into the matrix. The aggregates, in most cases, are pieces of marble, glass crystals, and crushed granite [3, 4]. According to some renamed companies, artificial stones have benefits that explain its importance to the consuming market. The main advantages are their solidity, good mechanical strength, and, in addition, their impermeability and low porosity. The low porosity is due to the use of the resin during the fabrication of the rocks, which interconnects the aggregates and penetrates between the interstices, thus eliminating the porosity of the artificial stone [4].

The articles of 'The world's largest iron exporter,' state that since April 2018 until now, Brazil has exported 25.2 million tons, showing the great economic importance that iron ore has for the country [5, 6]. However, the volume of waste generated is proportional to this great production. Landfills and dams are the usual destination for these materials, and, it might occupy a large area, contaminating the soil and the rivers, causing a series of problems for the environment and risks to human health [7].

This present work aims to develop an artificial rock with polymer matrix and disperse reinforcement of iron mineral ore to replace commercial rocks, and presents an alternative to the waste produced by the mining industries.

Materials and Methods

In this work, iron ore residue was heated in an oven at 100 °C for 24 h. The resin used was MC130, a diglycidyl ether of bisphenol A (DEGEBA) epoxy having a density of 1.15 g/cm^3 and the hardener, which is used for curing the resin, was FD 129, triethylenetetramine.

The plates confection process has begun by taking the residue to the oven at 100 °C for 24 h. Then, its mass was measured, and it was manually mixed with 15% by mass of the epoxy resin until it reaches homogeneity. The mixture was then added to the mold, having dimensions of 100 × 100 × 10 mm. After mold closure, it was taken to a 30 tons Schwing® press where it was pressed with a 10T load. The system remained on the slab for 2 h, then it was removed and maintained on a completely flat surface for 24 h at room temperature to continue the curing process. At the end of this period, the slabs were placed in the stove for 5 h and then passed to the characterized stage.

The density, water absorption, and porosity were determined following the guidelines of ABNT NBR 15845 (2010)—Annex B [8]; for the tests, 10 specimens were prepared.

The resistance to chemical attacks for artificial stone was determined using an adaptation of Annex H of ABNT NBR 13.818 [9]. The specimens were kept in a stove to dry and they were weighed before and after the chemical attack.

Thermogravimetric (TG) analysis was performed with the TGA—Q5000 TA Instrument. This analysis was carried out in a temperature range of 30–1000 °C with a heating rate of 10 °C/m, using an air flow of 60 ml/m. The thermogravimetric analysis was performed to determine the actual residue load in relation to the resin contained in the artificial stone.

The environmental analysis of the residue was performed through the leaching and solubilization tests, and it was possible to evaluate the hazardousness of the residue in relation to the metals. The abovementioned tests were carried out at the Center for Mineral Technology—CETEM, the reference standards were NBR 10004—Solid Waste Classification [10], NBR 10005—Leaching Test [11] and NBR 10006—Solubilization assay [12].

Results and Discussion

Table 1 shows the values related to density, apparent porosity, and water absorption of the artificial stone with iron ore residue and 15% epoxy resin (ASIO 15%). The average value found for an apparent density was 3.09 ± 0.01 g/cm^3, with the value above the commercial artificial marbles, which varies between 2.4 and 2.5 g/cm^3 [13]. Demartini et al. [14] found a value equal to 2.10 g/cm^3. When comparing to the ASIO 15%, it can be noticed that it has stood up, resulting in a denser, heavier material.

The average porosity found was $0.65 \pm 0.03\%$, and Chiodi and Rodriguez [15] have stated that the stones with porosity values below 0.5% are classified as a high-quality material for a civil construction. A value close to that found by Carvalho et al. [13] was found, with an average porosity of 0.55%.

In relation to the water absorption, the value of 0.21%, compared with the value of 0.25% found by Carvalho et al. [13], the ASIO 15% has shown less absorption. In addition, the water absorption falls between 0.009 and 0.40%, values presented by the industries producing artificial stone [15].

Table 1 Density, porosity, and water absorption of ASIO 15%

	ASIO 15%
Density (g/cm^3)	3.09 ± 0.01
Porosity (%)	0.65 ± 0.03
Water absorption (%)	0.21 ± 0.06

Table 2 Mass loss of ASIO 15%

Reagents	Mass lost (g)
Potassium hydroxide	0.014 ± 0.03
Hydrochloric acid	0.025 ± 0.07
Ammonium chloride	0.014 ± 0.05

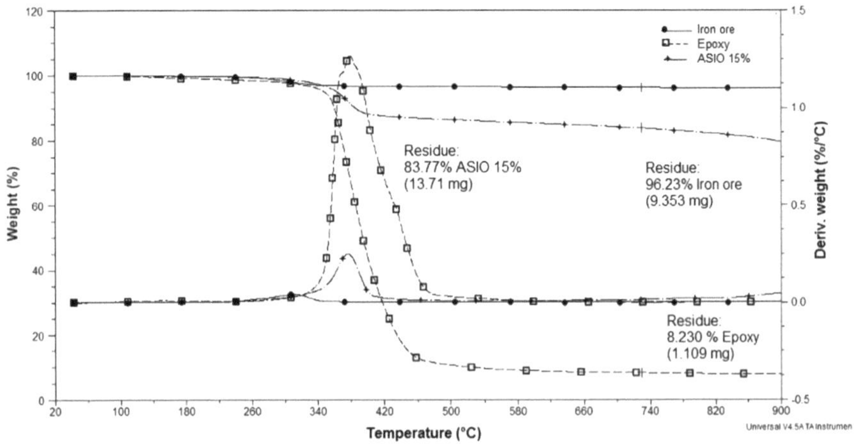

Fig. 1 Thermogravimetric analysis for the epoxy resin, the iron ore residue, and the ASIO 15%

The ability of a material to keep its surface intact after contact with certain substances and products determines its resistance to chemical attack. In Table 2, it is possible to observe the relative mass loss of ASIO 15% against the chemical attacks of potassium hydroxide, hydrochloric acid, and ammonium chloride. These reagents are present in daily life, potassium hydroxide is found in soaps, ammonium chloride is found in the composition of detergents and soaps and hydrochloric acid commercially known as muriatic acid is used as a cleaner [16].

Observing the results, it can be verified that the ASIO 15% lost mass when in contact with all the solutions used, however the most expressive loss occurred in contact with the hydrochloric acid, with loss of 0.025 g. Therefore, it is recommended to avoid the use of hydrochloric acid and do the cleaning only with the use of a damp cloth with neutral soap.

The graph of Fig. 1 shows the mass losses from the decomposition of the DGEBA–TETA resin, the iron ore residue, and the artificial stone produced in this project.

The iron ore residue is predominantly composed of kaolinite, goethite, hematite, gibbsite, and quartz [17]. The curve representing the iron ore residue did not show significant degradation in the test temperature range, and the small curve drop around 350 °C may be due to water loss of the goethite composition [18].

In the curve representing the epoxy resin, there is a mass loss of approximately 70% around 380 °C, due to resin degradation [4]. Only one residue of non-volatile material corresponding to an actual mass of 1.109 mg was left.

Table 3 Analytical results of the leaching extract

Parameters	Unit	Analytical results	MLA NBR 10004:2004
Arsenic	mg/L	<0.02	1.0
Barium	mg/L	0.32	70.0
Cadmium	mg/L	<0.001	0.5
Lead	mg/L	<0.02	1.0
Chrome	mg/L	0.12	5.0
Mercury	mg/L	<0.01	0.1
Silver	mg/L	<0.001	5.0
Selenium	mg/L	<0.02	1.0

Since that the iron ore residue did not present significant loss of mass, the loss of mass around 15% at about 380 °C at the curve representing the ASIO is associated with the resin, proving that the composition of the artificial stone was around 85% of residue and 15% of resin mass.

The leaching results for metals are presented in Table 3, when comparing these results with the maximum values allowed by NBR 10004—Annex F [10], it is verified that the obtained parameters did not exceed the allowed limits, thus, the residue does not present toxicity. Therefore, the iron ore residue in relation to the analyzed metals is classified as being class II, that is, non-hazardous waste.

The results of the solubilization test of the iron ore residue are presented in Table 4, comparing these results with the maximum values allowed by NBR 10004—Annex G [10], it is verified that the parameters arsenic, lead, and selenium present values above the maximum limits allowed. Thus, the iron ore residue is classified as non-inert—Class II A in relation to the metals analyzed.

Conclusions

An artificial stone with 85% of iron ore residue incorporated in 15% of DGEBA–TETA epoxy resin was developed. The iron ore residue from the environmental analysis was classified as non-inert—Class II A.

The absorption of water was 0.21% and this figure was between what the producers of artificial marble present in their material. Yet, the porosity was slightly above the value considered for quality materials applied in civil construction.

All solutions that came in contact with ASIO 15% promoted mass loss to a greater or lesser degree. Since hydrochloric acid promoted the loss of a more pronounced mass of 0.025 g, it is recommended that cleaning be carried out only with the use of a damp cloth with neutral soap.

The composition of the artificial stone was measured by thermogravimetric analysis as about 85% iron ore residue and 15% epoxy resin.

Table 4 Analytical results of the solubilization extract

Parameters	Unit	Analytical results	NBR 10004:2004
Arsenic	mg/L	<0.02	0.01
Barium	mg/L	0.12	0.7
Cadmium	mg/L	<0.001	0.005
Lead	mg/L	<0.02	0.01
Chrome	mg/L	<0.001	0.05
Mercury	mg/L	<0.001	0.001
Silver	mg/L	<0.001	0.05
Selenium	mg/L	<0.02	0.01
Aluminum	mg/L	<0.04	0.2
Copper	mg/L	0.02	2.0
Iron	mg/L	<0.001	0.3
Manganese	mg/L	<0.002	0.1
Sodium	mg/L	4.7	200
Zinc	mg/L	0.16	5.0

The production of artificial stone using iron ore residue for the production of lining in the civil construction area was proved as an alternative to give a sustainable destination to the waste and thus add value to it.

Acknowledgements The authors thank FAPERJ and CNPq for the financial support, the CETEM and the technician Renan da Silva Guimarães for the support during the experiments.

References

1. Ribeiro CEG, Rodriguez RJS (2015) Influence of compaction pressure and particle content on thermal and mechanical behavior of artificial marbles with marble waste and unsaturated polyester. Mater Res. https://doi.org/10.1590/1516-1439.372314
2. Brazilian Association of the Ornamental Rocks Industry-ABIROCHAS (2018, in Portuguese) Balance of Brazilian exports and imports of ornamental stones in 2017, Brasília
3. Ribeiro CEG, Rodriguez RJS, Vieira CMF, Carvalho EA, Candido VS, Monteiro SN (2014) Production of synthetic ornamental marble as a marble waste added polyester composite. Mater Sci Forum 775–776:341–345. https://doi.org/10.4028/www.scientific.net/MSF.775-776.341
4. Silva FS, Ribeiro CEG, Rodriguez RJS (2018) Physical and mechanical characterization of artificial stone with marble calcite waste and epoxy resin. Mater Res. http://dx.doi.org/10.1590/1980-5373-MR-2016-0377
5. Brazilian Mining Association–IBRAM (2018, in Portuguese) Annual activity report - June 2017–June 2018, Brasília
6. Institute of Applied Economic Research–IPEA (2012, in Portuguese) Diagnosis of solid waste from mining activity of non-energy substances, Brasília
7. Brazilian Mining Association–IBRAM (2016, in Portuguese) Management of mining tailings, Brasília

8. Brazilian Association of Technical Norms–ABNT (2010, in Portuguese) NBR 15845: Coating stone - test methods, Rio de Janeiro
9. Brazilian Association of Technical Norms–ABNT (1997, in Portuguese) NBR 13818: ceramic tiles - specification and methods of test, Rio de Janeiro
10. Brazilian Association of Technical Norms–ABNT (2004) NBR 10004: solid waste – classification, Rio de Janeiro
11. Brazilian Association of Technical Norms–ABNT (2004) NBR 10005: procedure for obtaining leach extract from solid wastes, Rio de Janeiro
12. Brazilian Association of Technical Norms–(2004) NBR 10006: Procedure for obtaining solubilized solid waste extract, Rio de Janeiro
13. Carvalho EAS, Vilela NFV, Monteiro SN, Vieira CMF, Silva LC (2018) Novel artificial ornamental stone developed with quarry waste in epoxy composite. Mater Res. http://dx.doi.org/10.1590/1980-5373-MR-2017-1104
14. Demartini TJ, Rodríguez RJS, Silva FS (2018) Physical and mechanical evaluation of artificial marble produced with dolomitic marble residue processed by diamond-plated bladed gang-saws. J Mater Res Technol. https://doi.org/10.1016/j.jmrt.2018.02.001
15. Chiodi CF, Rodriguez H (2009) Guide application of stone coverings. ABIROCHAS, São Paulo, Brazil, p 118
16. Bolonini TM, Godoy AM (2012, in Portuguese) Technological characterization of the ornamental stone of the Maciço Capão Bonito, Geociências 31:229–246
17. Almeida CA, Oliveira AF, Pacheco AA, Lopes RP, Neves AA, Queiroz MELR (2018) Characterization and evaluation of sorption potential of the iron mine waste after Samarco dam disaster in Doce River basin – Brazil. Chemosphere 209:411–420
18. Saito G, Nomura T, Sakaguchi N (2016) Akiyama T (2016) Optimization of the dehydration temperature of goethite to control pore morphology. ISIJ Int 56(9):1598–1605

Study of the Incorporation of Waste from the Paper Industry in Ceramic Tiles

A. R. G. Azevedo, B. C. Mendes, M. T. Marvila, J. Alexandre, E. B. Zanelato, G. C. Xavier, N. A. Cerqueira, S. N. Monteiro and T. E. S. Lima

Abstract The incorporation of waste from industries in construction is becoming an increasingly common practice; since in addition to decrease the amount of material disposed on the environment, it can further improve its properties. Among the residues in need of disposal is the waste from the paper industry. The objective of this work is to evaluate the incorporation of paper waste in ceramic tile. Thus, 0, 10, 20, and 30% incorporations were used to compare the results. The materials used in the research, clay and residue, were characterized by the grain size tests, chemical analysis, and actual grain density. To evaluate the best content of incorporation were made specimens of test and realized the tests of resistance to flexion, firing linear shrinkage and water absorption. The results indicated that it is feasible to use the incorporation of 10% residue, only because of economic and environmental issues, since the properties are altered in this incorporation.

Keywords Waste · Paper industry · Ceramic tiles · Sustainability

A. R. G. Azevedo
TER—Department of Agricultural Engineering and Environment, UFF—Federal Fluminense University, Rua Passo da Pátria, 341, Niterói, Rio de Janeiro 24210240, Brazil

A. R. G. Azevedo (✉) · E. B. Zanelato
IFF—Federal Institute Fluminese, DIRINF—Directorate of Infrastructure Rectory, Rua Cel. Valter Kramer, 357 - Parque Vera Cruz, Campos dos Goytacazes, Rio de Janeiro 28080-565, Brazil
e-mail: afonso.garcez91@gmail.com

B. C. Mendes
UFV—Federal University of Viçosa, DEC, Av. Peter Henry Rolfs, s/n - Campus University, Viçosa, Minas Gerais 36570-000, Brazil

M. T. Marvila · J. Alexandre · E. B. Zanelato · G. C. Xavier · T. E. S. Lima
LECIV—Civil Engineering Laboratory, UENF—State University of the Northern Rio de Janeiro, Av. Alberto Lamego, 2000, Campos dos Goytacazes, Rio de Janeiro 28013-602, Brazil

N. A. Cerqueira
UNIREDENTOR—Redentor University Center, BR-356, 25 - Pres. Costa e Silva, Itaperuna, Rio de Janeiro 28300-000, Brazil

S. N. Monteiro
Department of Materials Science, IME—Military Institute of Engineering, Square General Tibúrcio, 80, Rio de Janeiro 22290-270, Brazil

S. Ikhmayies et al. (eds.), *Green Materials Engineering*, The Minerals, Metals & Materials Series, https://doi.org/10.1007/978-3-030-10383-5_30

Introduction

Industrial activities are responsible for the generation of large quantities of solid waste, which are often disposed of or stored in the environment in inadequate places or without any previous treatment. This form of disposal causes several economic and environmental impacts, since companies need to invest a considerable amount to acquire land for the allocation of these by-products. In relation to the environment, waste can lead to contamination of water, air, and soil [1, 2].

The paper industry, for example, achieved a production of 401 million tonnes by the year 2015, given the global context [3]. In Brazil, this number corresponds to 10.4 million tons. Associated with this, high production was the generation of approximately 1.3 million tons of by-products in the same year [4]. Given the large volume of waste resulting from the papermaking stages, alternatives should be studied for reuse and recycling, turning these discarded materials into potential raw materials.

One possible solution is the incorporation of industrial waste into building materials, which has become a common practice in recent years. Several studies have been carried out in this field [5, 8] addressing the most varied types of by-products, such as fly ash, mining tailings, sand dust, and limestone powder. In this way, the volume of waste disposed in the environment is reduced and, at the same time, value is added to the material, which can be applied as a new draw. In addition, the properties of the residue can be improved and potentiated according to the desired use.

Waste from the paper industry has been studied and applied to cement-based products for civil construction, such as concrete [9], mortars [10] and cements [11]. Another option is to use the waste as a raw material in the ceramics industry. In general, this material presents properties that favor the processes of sintering and drying [12], justifying its use in the production of ceramic pieces. On the other hand, the ceramic industry is strongly promising in the absorption of discarded materials, due to the burning step in the manufacturing process. Thus, compounds that may exhibit some toxicity can be eliminated by means of volatilization, chemical reactions, or immobilization in the vitreous phase form [13].

The present work has an objective to evaluate the effects of the incorporation of waste from the paper industry in ceramic tiles. Binary mixtures composed of clay and the residue were prepared, with contents of 0, 10, 20, and 30% of residue by mass. The properties of tensile strength in the bending, water absorption, and linear retraction in the burning were evaluated, in order to provide the technical feasibility of addition of the residue, in accordance with the criteria specified in the current regulations. In this way, there is a promising alternative not only from the environmental point of view, but also guarantees quality to the final product.

Materials and Methods

The clay used in this work comes from a deposit located in the municipality of Campos dos Goytacazes, Brazil. The waste from the paper industry was collected at COPAPA, located in the northwestern part of the state of Rio de Janeiro, Brazil. The materials were transported to the Civil Engineering Laboratory (LECIV)—UENF, where they were allocated and prepared for the characterization and production tests of the ceramic pieces.

The physical and chemical characterization of the raw materials was carried out. For the physical characterization, the granulometric analysis and specific mass of the grains of the clay were carried out, following the requirements of NBR 7181 [14] and NBR 6508 [15], respectively. In the determination of the granulometric distribution, the fraction retained in the ASTM 200 mesh sieve was classified by sieving; the bypass fraction, in turn, was classified by sedimentation. The Casagrande standard—Unified Soil Classification System—was used to classify materials. For the chemical characterization, the test was performed to determine the chemical composition of the raw materials, allowing the identification and quantification of the oxides that constitute them; this procedure was performed for both the clay and the residue. The technique of energy-dispersive X-ray spectroscopy (EDX) was used with the aid of the Shimadzu EDX 700 equipment.

After the characterization of the materials, the ceramic masses were prepared, with different levels of incorporation of the residue: 0, 10, 20, and 30%. Then, prismatic specimens were produced by extrusion with the aid of a laboratory extruder. The specimens were air-dried and then heated to 750 °C. After cooling, the following characterization tests were performed: linear variation in the burning, apparent porosity, water absorption, and flexural strength at three points.

Based on the results obtained, the best level of incorporation of the residue was selected. Then, tiles were produced with the residue content defined and without addition (reference standard). The firing of the tiles was done in a similar manner to that of the specimens, in the same oven. After burning, the parts were characterized according to the procedures prescribed in the standard NBR 15310-1 [16].

Result and Discussion

For the characterization, it was initially adopted the determination of the granulometric fractions of the soil used, as can be observed in Table 1.

The soil characterization, shown in Table 1, presents a grain size distribution of 57% of clay and 26.5% of silt, accounting for 83.5% of the material only in these two particle sizes; however, the paper industry residue presents 37.8% of clay and 50.8% of silt and totaling 88.6% of these two bands; this proximity between the bands shows a possible favorable interaction in the development of mixtures and both can be determined as silt–clay material [17]. The soil had a true grain density

Table 1 Sizing characteristics of the soil used

Diameter of the Grains (mm)–ABNT NBR 6502: 1995–Rocks and Soils							
Clay	Silt	Sand			Boulder		
		Thin	Medium	Thick	Thin	Medium	Thick
<0.002	0.002–0.06	0.06–0.2	0.2–0.6	0.6–2	2–6	6–20	20–60

Granulometric Features							
Boulder			Sand			Silt	Clay
Thick	Medium	Thin	Thick	Medium	Thin		
–	–	–	–	2.2%	3.3%	26.5%	57.0%

Table 2 Chemical characterization by EDX of the paper manufacturing waste in percentage of the compound

	CaO	Al_2O_3	SiO_2	K_2O	SO_3	Fe_2O_3	TiO_2	SrO	ZnO	ZrO_2
Soil	1.00	35.87	45.32	2.15	1.31	11.81	1.72	–	–	–
Residue	79.006	8.898	8.559	1.163	1.106	0.751	0.352	0.105	0.052	0.007

of 2.63 g/cm^3 while the residue had a density of 1.95 g/cm^3 and a mean moisture content of 55%. Table 2 shows the chemical characterization.

The specific mass of the residue is smaller than that of the small aggregate used, this lower specific mass is due to the fact that larger internal volumes of the unfilled material and their constitution, which can be confirmed by the chemical analyzes, is shown in Table 2. The results presented in Table 2 show that the residue consists predominantly of calcium oxide (CaO), silica (SiO_2), and alumina (Al_2O_3). CaO is associated with calcite. Already SiO_2 and Al_2O_3 are combined forming the kaolinite structure. Carbon is associated with the presence of cellulose. The remaining oxides present in TiO_2 and Fe_2O_3 are possibly impurities that present in kaolin and calcite [10]. The soil has a silicon oxide composition which is a strong indicator of the presence of free silica and clay minerals, such as kaolinite ($Al_2O_3 \bullet 2SiO_2 \bullet 2H_2O$), clay mineral commonly verified by researchers in the region [18].

As described in the methodology, tests were carried out on prismatic specimens in the proportions of 0, 10, 20, and 30% of waste incorporation in the ceramic mass, making it possible to determine the best percentage to be used in making the tiles in real size; thus, Fig. 1 shows the linear variation (%) of the specimens.

It is observed in Fig. 1 that the increase in the percentage of inclusion reduces the linear variation caused in the burning process, and this is due to the transformations that occur within the ceramic matrix, which when mixed with the paper residue produces compounds that fill the pores initially reducing the lien imposition imposed [10]. Figure 2 shows the behavior of water absorption (%) in prismatic specimens.

It is observed in Fig. 2 that even with the reduction of linear variation, observed in Fig. 1, the water absorption increases as the incorporation of the residues increases, this is due to the large porosity present in this type of material, which even benefited for its use in the matrix ceramics still exhibits a porous behavior, contributing to an

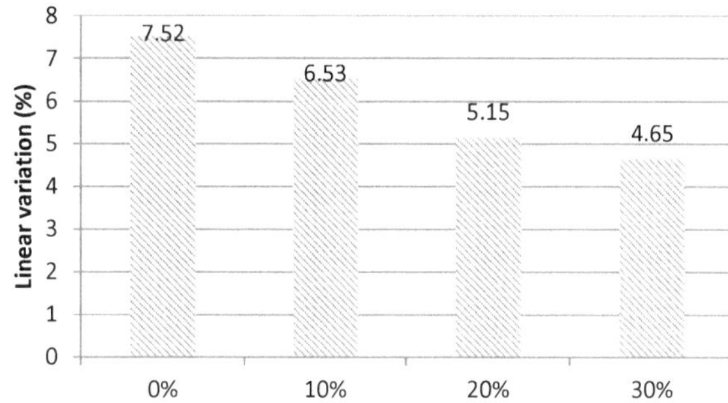

Fig. 1 Influence of the percentage of incorporation in linear retraction (%)

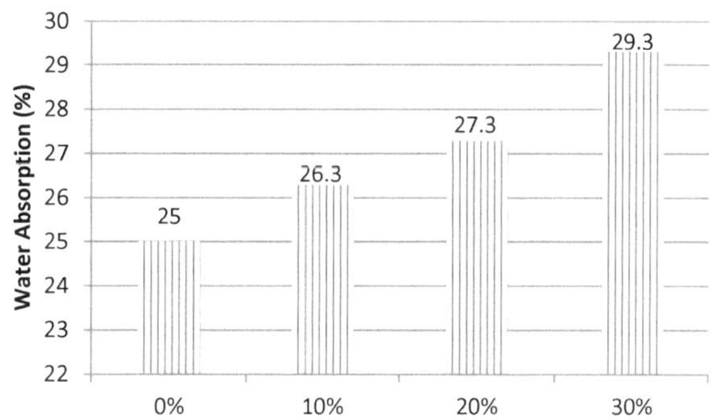

Fig. 2 Influence of the percentage of incorporation in water absorption (%)

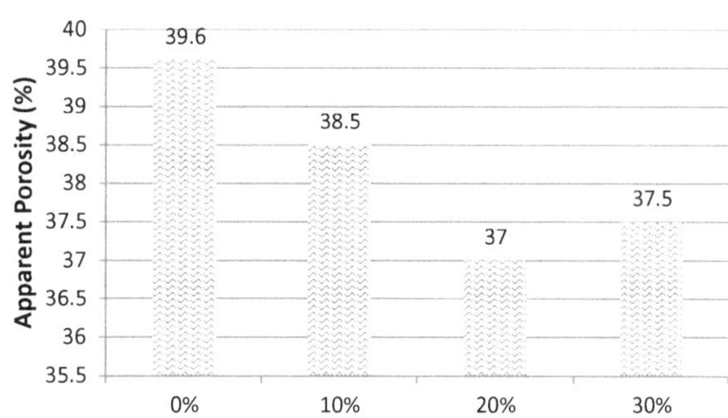

Fig. 3 Influence of the percentage of incorporation in apparent porosity (%)

increase in global water absorption [19]. Figure 3 shows the apparent porosity (%) at the different levels of incorporation, and Fig. 4 is the mechanical resistance to bending.

It is observed in Fig. 3 that the increase in the percentage of incorporation significantly reduced the apparent porosity; mainly at 20% levels, this fact is due to the superficial filling that the residue, after passing through the burning process of the ceramic material, this effect has already been described in the literature as being an important agent favoring the incorporation of these materials [1]. However, very low values can be harmful, such as those presented in 20%. As for the mechanical resistance, the incorporation of 10% showed an improvement in this property, and it

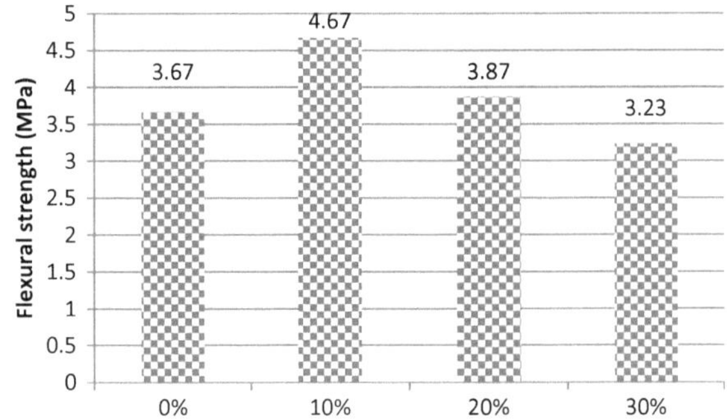

Fig. 4 Influence of the percentage of incorporation in flexural strength (MPa)

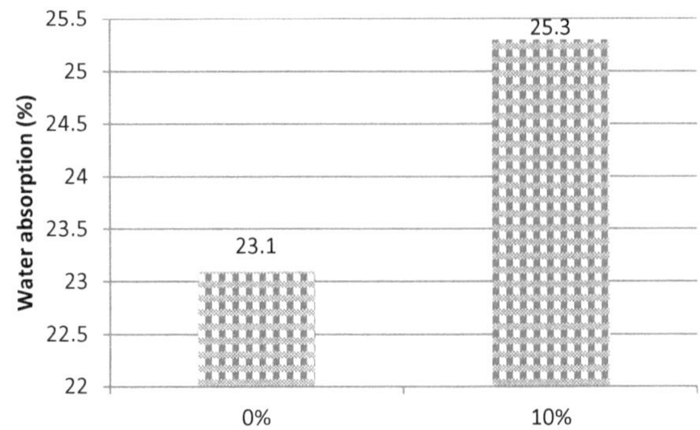

Fig. 5 Water absorption in the traits evaluated (%)

was not verified in a linear trend, since the transformations that occur after the firing stage significantly reduces this parameter [10, 20].

Thus, after analyzing the results of the specimens shown in Figs. 1, 2, 3, and 4, it can be concluded that the mixture with 10% is the one that meets the main requirements and has economic and environmental feasibility, which is used to make the tiles in real size.

A total of 20 tiles were molded for each batch, one of 0% and another of 10%, for each of the tiles, each of which was divided for water absorption, mechanical strength and waterproofing tests. Regarding impermeability, the test consists in verifying the formation of a trace of moisture on the underside of the tile, when subjected to a water accumulation condition. Under these conditions, both traces (0 and 10%) were considered impermeable. Figures 5 and 6 show the mean water absorption and mechanical strength results, respectively, of the traits evaluated.

The results of Figs. 5 and 6 show that the use of pulp and paper industry waste is viable in the production of tiles, in the proportion of 10%, even incurring a reduction in resistance, which is within the limits suggested by the literature [21]. The resistance reduction characteristics arise due to the reactions that the compounds of the paper residue release with the burning process together with the clay material, and this process does not interfere in the resistance in a significant way in order to make it unfeasible, according to reports in the literature [12].

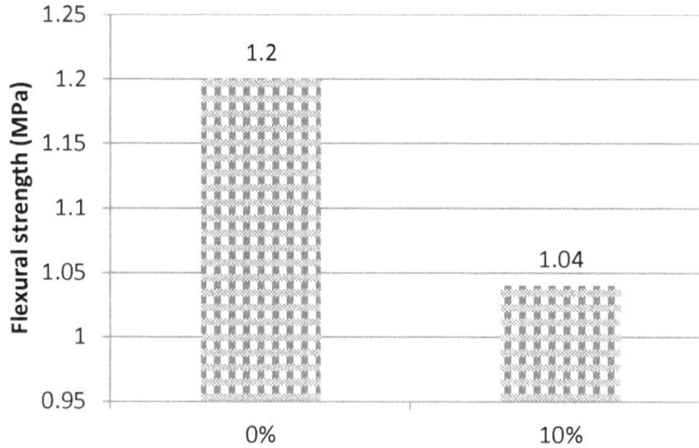

Fig. 6 Mechanical resistance to flexion of the traces evaluated (MPa)

Conclusions

With the results observed in this research, we can indicate that:

- The use of waste from the pulp and paper industry is economically and environmentally efficient, providing a cost reduction in the disposal process of this waste within Brazilian environmental standards;
- As for the characterization of the material, the same was shown within the data available in the literature, and its interaction within the ceramic matrix plays a satisfactory role suggesting the feasibility of its use;
- The incorporation of 10% of the residue in the clay mass was technically feasible, as in the issue of impermeability, absorption and water and mechanical resistance, which suggests that the construction of these ceramic artifacts is advantageous.

Acknowledgements The authors thank the Brazilian agencies: CNPq, CAPES, and FAPERJ for the support provided to this investigation.

References

1. Vieira CMF, Pinheiro RM, Rodriguez RJS, Candido VS, Monteiro SN (2016) Clay bricks added with effluent sludge from paper industry: technical, economical and environmental benefits. Appl Clay Sci 132:753–759
2. Sutcu M, Díaz JJC, Rabanal FPA, Gencel O, Akkurt S (2014) Thermal performance optimization of hollow clay bricks made up of paper waste. Energy Build 75:96–108
3. BRACELPA—Brazilian Pulp Association (2016, in Portuguese). Annual publication of the Brazilian paper and pulp industry. Accessed 29 June 2018
4. Federation of Industries of the State of Paraná (2017, in Portuguese) Annual report, 1 edn. FIEP, Curitiba
5. Chen Y, Zhang Y, Chen T, Zhao Y, Bao S (2011) Preparation of eco-friendly construction bricks from hematite tailings. Constr Build Mater 25:2107–2111
6. Kute S, Deodhar SV (2003) Effect of fly ash and temperature on properties of burnt clay bricks. J Civ Eng 84:82–85

7. Coronado M, Segadães AM, Andrés A (2014) Combining mixture design of experiments with phase diagrams in the evaluation of structural ceramics containing foundry by-products. Appl Clay Sci 101:390–400
8. Bentz DP, Ferraris CF, Jones SZ, Lootens D, Zunino F (2017) Limestone and silica powder replacements for cement: early-age performance. Cement Concr Compos 78:43–56
9. Ahmadi B, Al-Khaja W (2001) Utilization of paper waste sludge in the building construction industry. Resour Conserv Recycl 32:105–113
10. Azevedo ARG, Alexandre J, Xavier GC, Pedroti LG (2018) Recycling paper industry effluent sludge for use in mortars: a sustainability perspective. J Clean Prod 193:335–346
11. Frias M, Rodriguez O, Sanchez de Rojas MI (2015) Paper sludge, an environmentally sound alternative source of MK-based cementitious materials. A review. Constr Build Mater 74:37–48
12. Cusidó JA, Cremades LV, Soriano C, Devant M (2015) Incorporation of paper sludge in clay brick formulation: Ten years of industrial experience. Appl Clay Sci 108:191–198
13. Pinatti DG, Conte RA, Borlini MC, Santos BC, Oliveira I, Vieira CMF, Monteiro SN (2006) Incorporation of the ash from cellulignin into vitrified ceramic tiles. J Eur Ceram Soc 26:305–310
14. Brazilian Association of Technical Norms (1984, in Portuguese) NBR 7181. Soil - Grain size analysis, Rio de Janeiro
15. Brazilian Association of Technical Norms (1984, in Portuguese) NBR 6508. Soil grains that pass through the sieve of 4.8 mm—determination of the specific mass, Rio de Janeiro
16. Brazilian Association of Technical Norms (2009, in Portuguese) NBR 15310. Ceramic components—tiles—terminology, requirements and test methods, Rio de Janeiro
17. Azevedo ARG, França BR, Alexandre J, Marvila MT, Zanelato EB, Xavier GC (2018) Influence of sintering temperature of a ceramic substrate in mortar adhesion for civil construction. J Build Eng 19:342–348. https://doi.org/10.1016/j.jobe.2018.05.026
18. Vieira, CMF, Souza, ETA, Monteiro SN (2004, in Portuguese) Effect of the incorporation of chamote in the processing and microstructure of red ceramics. Cerâmica, São Paulo 50 (315):254–260
19. Pinheiro RM, Vieira CMF, Sánchez RJ, Monteiro SN (2008, in Portuguese) Recycling of waste from paper production in red ceramic. Revista Matéria 13(1):220–227
20. Simão L, Jiusti J, Lóh NJ, Hotza D, Raupp-Pereira F, Labrincha JA, Montedo ORK (2017) Waste-containing clinkers: valorization of alternative mineral sources from pulp and paper mills. Process Saf Environ Prot 109:106–116
21. Buruberri LH, Seabra MO, Labrincha JA (2015) Preparation of clinker from pulp industry wastes. J Hazard Mater 286:252–260

Study of the Technological Properties of Multiple Mortar Use with Efficient Addition of Rock Waste

L. F. Amaral, M. Nicolite, G. C. G. Delaqua, M. T. Marvila, J. Alexandre, S. N. Monteiro and Carlos Mauricio Fontes Vieira

Abstract The search for ideal mixture of aggregates in order to increase mortar density has become an interesting topic to improve mortar properties. During ornamental rocks processing, about 25–30% of the blocks are transformed into powder as waste. Under this situation, this work proposes the rock waste incorporation as partial replacement of sand by using Alfred's mathematical model to calculate the ideal granulometric distribution. The elaborated mortars are based on Portland cement CPII E 32, hydrated lime CHIII, sand, and residue from the processing of ornamental rocks, denominated as rock waste. The raw materials were characterized in terms of granulometric distribution. It can be concluded the efficiency in used ornamental rock waste and mathematical model for granulometric optimization to improve mortar density.

L. F. Amaral (✉) · M. Nicolite · G. C. G. Delaqua · M. T. Marvila · J. Alexandre · C. M. Fontes Vieira
Advanced Materials Laboratory—Science and Technology Center State University of North Fluminense Darcy Ribeiro, Alberto Lamego Avenue, 2000, Campos dos Goytacazes 28013-602, Brazil
e-mail: lfamaral28@gmail.com

M. Nicolite
e-mail: micaela_uenf@yahoo.com.br

G. C. G. Delaqua
e-mail: geovanagirondi@gmail.com

M. T. Marvila
e-mail: markssuel@hotmail.com.br

J. Alexandre
e-mail: jonasuenf@gmail.com

C. M. Fontes Vieira
e-mail: carlos.vieira@pq.cnpq.br

S. N. Monteiro
IME (Military Institute of Engineering), Praça General Tibúrcio 80, Praia Vermelha, Urca, Rio de Janeiro, RJ 22290-270, Brazil
e-mail: snevesmonteiro@gmail.com

S. Ikhmayies et al. (eds.), *Green Materials Engineering*, The Minerals, Metals & Materials Series, https://doi.org/10.1007/978-3-030-10383-5_31

Keywords Packaging · Density · Granulometric distribution · Waste incorporation

Introduction

Mechanical properties and durability of concrete are highly affected by aggregate properties such as aggregate gradation [1–8]. And this can be extended to mortars as Paiva (2015) proposes [9], in which good aggregate gradation corresponds to high packing density and properties of mortars and concrete. In this way, the search for the ideal mixture of aggregates in order to increase its compactness becomes an interesting proposal in terms of improvements in the properties of the mortar.

Some researches [10–12] developed mathematical methods to calculate granulometric distributions that could be used for greater packing of the aggregates, particularly in concretes.

Among these methods are two classical approaches, one discrete, which considers the particles individually, known as Furnas model [13] and the other continuous one, proposed by Andreasen and Andersen [14], known as Andreasen model. However, Funk and Dinger [15] have proved that the Furnas and Andreasen models converge mathematically to the same equation, known as Alfred's model, when both equations are calculated with infinitely small particles. Thus, Alfred's equation is currently considered the most suitable particle packaging model for real systems.

In this context, it was verified that residues from the processing of ornamental rocks could increase the density of mortars by means of their insertion in the mixture of aggregates [10].

In the municipality of Santo Antônio de Pádua, located 150 km from Campos dos Goytacazes, RJ, there is intense industrial activity related to the extraction and processing of ornamental rocks. During processing these rocks, about 25–30% of the blocks are transformed into powder, which is estimated to be generated in Brazil 240,000 ton/year of these waste rocks [16]. In this way, the correct destination of this waste has been of concern for the companies, given its costly cost. In view of this situation, studies that point out definitive solutions in terms of reuse of this residues and that possess economic viability become great allies in the practice of sustainable development.

This work proposes the incorporation of rock waste as partial replacement of sand by using Alfred's mathematical model to calculate the ideal granulometric distribution to provide the greater packaging.

In this way, besides the possibility of increasing the performance of mortars, this work seeks to minimize the environmental impact caused by the generation of this residue, as well as to propose alternatives for the cost reduction in the attempt of its correct destination, proposing to replace part of the sand of the mortars multiple use by ornamental rock waste.

Materials and Methods

Materials

The elaborated mortars in this work are based on cement (**C**), hydrated lime (**HL**), sand (**S**), and residue from the processing of ornamental rocks, denominated as rock waste (**RW**).

Portland cement CPII E 32 Votoram was used as binder, and the natural quartz sand was used as small aggregate and comes from the Paraíba do Sul River, located in the municipality of Campos dos Goytacazes—RJ, and the hydrated lime adopted was type CHIII, of the Supercal brand. These materials were chosen because they are commonly used in the preparation of mortars in Campos dos Goytacazes. The rock waste comes from an ornamental rock charcoal industry located in Santo Antônio de Pádua—RJ and was collected following solid waste sampling procedures.

Methods

The raw materials were characterized according to their specific mass, granulometric distribution, and chemical composition. The determination of sand and waste chemical composition, in the form of oxides, was obtained by dispersive energy X-ray fluorescence (EDX) technique.

Mortar Preparation

All the produced mortars in this work were guided by Brazilian standard. The mixing was performed in a mechanical mixer provided with a steel vat having a capacity of approximately five liters and having a metal blade rotating about itself in planetary motion about the axis of the vat.

Raw Materials Dosing

Alfred's mathematical model is prescribed by a continuous distribution and takes into account the diameter of the smallest particle (Ds) and the largest particle (Dl) of the mixture. This equation is described by:

$$CPFT(\%) = 100\,x\left(\frac{D^q - D_S^q}{D_L^q - D_S^q}\right) \quad (1)$$

whereby CPFT represents the cumulative percentage of particles having sizes smaller than D. The distribution modulus q influences the ratio between coarse and fine particles.

Higher values of the distribution modulus ($q > 0.5$) are leading to coarse mixtures, whereas smaller values ($q < 0.25$) are resulting in mixtures which are rich in fine particles [10]. As will be seen, the rock waste has its grain size distribution composed predominantly of small particles; thus, the value of 0.22 was the q value admitted to guarantee the higher mortar density. The q value equal to 0.37 was chosen because it represents the maximum theoretic density as proposed by Andreasen and Andersen [14].

Therefore, different grain size ranges of the raw materials that will be part of the aggregates of the mortar were pre-established. Then, through Alfred's model, the percentage volumes relative to each diameter range were calculated to reach the most compact structure.

After analyzing the granulometric distribution of each aggregate, the final granulometry of the mixture will be the weighted average of the distributions, according to Eq. 2, in which the multiplication factors (β_j) are the respective percentages of each aggregate in the mixture. Thus, the approximation of the final granulometry of the mixture to the granulometry proposed by the mathematical model was performed by the least squares method. This approach is done by means of the optimization obtaining the smallest sum of the square of the residuals (RSS) between the ideal granulometry and the granulometry of the mixture; in this process the Solver tool Excel software was used.

$$pc_i = \sum_{j=1}^{m} \beta_j pa_{j,i} \tag{2}$$

So, for n diametric ranges and m aggregates according to Table 1: pe_i = accumulated retained percentage specified for the sieve range "i"; $pa_{j,i}$ = retained percentage accumulated in the sieve range "i" for aggregate a_j; pc_i = retained percentage accumulated in sieve range "i" for the mixture of aggregates; Δ_i = deviation of the mixture from the specification in the "i" range; β_j = percentage of aggregate a_j in the combination; MD = maximum deviation, that is, the sum of the squares of the individual residuals. In that, $1 \leq i \geq n$ and $1 \leq j \geq m$. Thus, the optimization consists in varying the β_j coefficients to reach a minimum value for MD, in which $\sum_{j=1}^{m} \beta_j = 1$ act as the volumetric constraint; in other words, it is a curve fitting problem.

Thus, the experimental design was elaborated with three mixes (reference mix plus two mixes using the proposed aggregate mixing methodology) and six specimens for each population. For the tests in the fresh and hardened state, all the mortars produced followed the Brazilian standards, being evaluated the consistency index, incorporated air embedded and fresh body density. And for the hardened state, the bulk density, flexural strength, and compression strength were performed.

Table 1 Spreadsheet model used to calculate the aggregate mix

Φ	pe (%)	pa_1 (%)	pa_2 (%)	…	pa_m(%)	pc (%)	Δ
Φ_1	pe_1	$pa_{1,1}$	$pa_{2,1}$	…	$pa_{m,1}$	pc_1	$\Delta_1 = pc_1 - pe_1$
Φ_2	pe_2	$pa_{1,2}$	$pa_{2,2}$	…	$pa_{m,2}$	pc_2	$\Delta_2 = pc_2 - pe_2$
…	…	…	…	…	…	…	…
Φ_n	pe_n	$pa_{1,n}$	$pa_{2,n}$	…	$pa_{m,n}$	pc_n	$\Delta_n = pc_n - pe_n$
MD							$\Sigma\Delta_i^2$

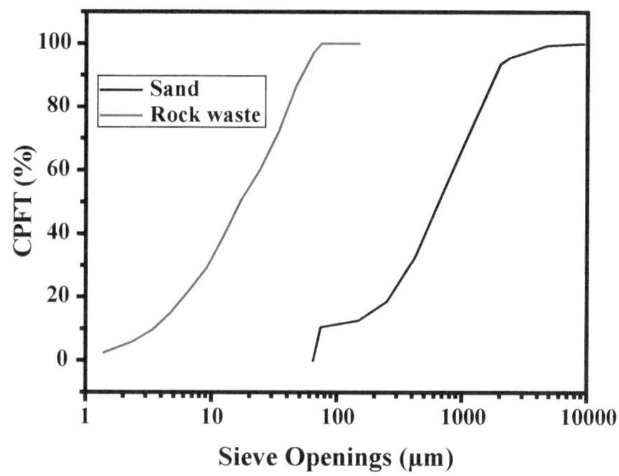

Fig. 1 Grain size distribution of sand and rock waste

Results and Discussion

The grain size distributions of the raw materials used as aggregate are illustrated in Fig. 1. It can be seen that the rock waste has a finer grain size distribution than sand. This is an interesting result, since the finer particles of the rock waste can fill the voids left by the sand particles accommodation increasing the mortar density.

Two mixes were made varying the distribution modulus coefficient (q), and this is represented in Figs. 2 and 3, for $q = 0.22$ and $q = 0.37$, respectively. It is observed that for $q = 0.37$ the theoretical grain size distribution curve demands a larger amount of coarse particles when compared with $q = 0.22$. Thus, smaller values for q provide the particle size distribution based on finer particles; thus, a greater amount of residue can be applied.

In Table 2, all the raw materials used to make the mortars are described in percentage weight. Three mixtures were prepared, and the amount of cement and hydrated lime were kept constant in all of them to not influence the differences of the evaluated properties. To evaluate the influence of the residue incorporation as an aggregate in mortars, 9% by weight of rock waste was incorporated for the formulation 91Ar and 21% for the 79Ar formulation, according to the results suggested by Alfred's mathematical model. Thus, to adjust the workability of the mortar the amount of

Fig. 2 Grain size distribution of Alfred's model for q = 0.22 and for the mixture (79Ar) that best approaches this distribution

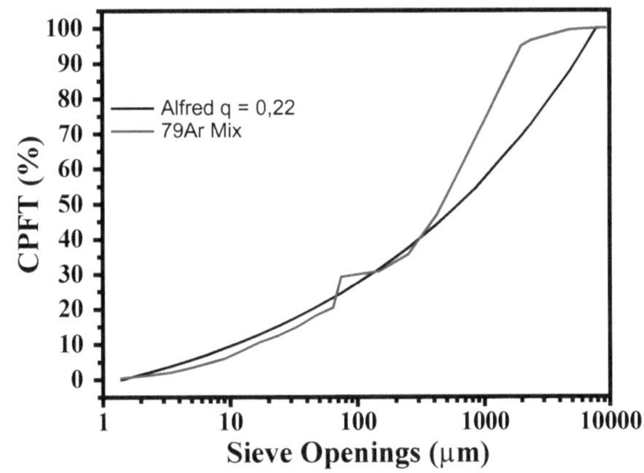

Fig. 3 Grain size distribution of Alfred's model for q = 0.37 and for the mixture (91Ar) that best approaches this distribution

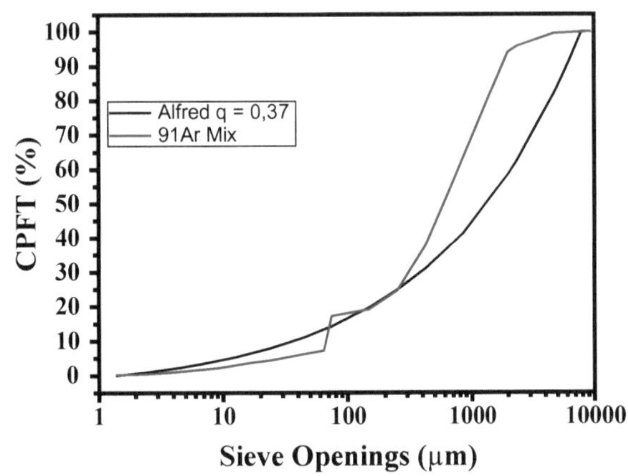

Table 2 Formulations in percentage of mass of the raw materials used in the preparation of mortar

No	q	C (kg/m^3)	HL (kg/m^3)	S (kg/m^3)	RW (kg/m^3)	W (kg/m^3)	W/C
100Ar	–	150	150	900	0	223	1.48
91Ar	0.37	150	150	819	91	213	1.42
79Ar	0.22	150	150	711	189	203	1.35

water had to be varied, where it is noticed that the increase in the rock waste content did made that the mortars required less water.

Traditionally, water/cement (w/c) ratio is probably the most important parameter within cement materials technology such as concrete and mortar. Although the increase in w/c ratio can increase the workability of mortar, results indicate that this increase reduces the values of hardened properties [17]. It is possible to notice that the water amount in the studied mixes was reduced as the increase of the residue incorporation; this reduction was necessary to maintain the same workability and

Table 3 Consistency index of the mixes mortars studied

Mixes	Consistency index (mm)	Water content (g)
100Ar	262	223
91Ar	259	213
79Ar	260	203

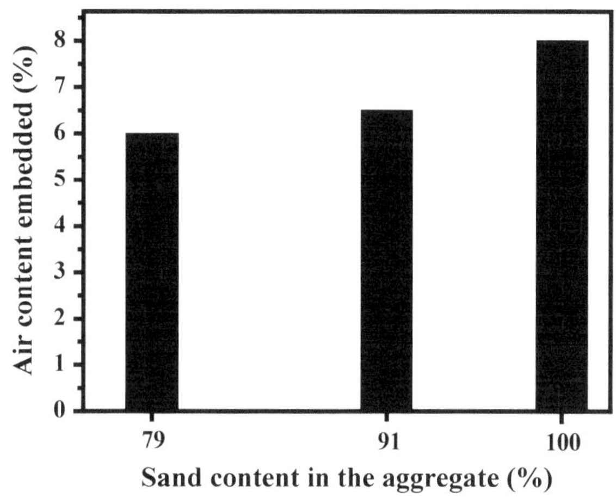

Fig. 4 Air content embedded of the mixes mortars studied

consistency of all mortar mixes. Thus, it is understood that the residue incorporation can reduce the void spaces increasing the mortar compactness, offering less space for water filling (Table 3).

The density and incorporated air embedded are inversely correlated properties; therefore, with increased density, there is a decrease in matrix voids, resulting in a lower content of air incorporated [18]. Therefore, as can be seen in Fig. 4, the 79Ar and 91Ar mixes were able to reduce the incorporated air embedded compared to the reference 100Ar mix. This may indicate the suitability of Alfred's model as a way of improving the efficiency of the mixture of rock waste as an aggregate in partial replacement of sand.

It can be seen in Fig. 5 that both formulations proposed by the mathematical model were able to increase the density in both fresh and hardened state of the mortars. It is also possible to note that the formulation 79Ar presented the highest value for this property, indicating that for the coefficient $q = 0.22$, which assumes the highest concentration of smaller particles in the mixture, it was more efficient when compared to the coefficient $q = 0.37$. Thus, the increase in density in the fresh state will positively influence the other evaluated properties, as will be seen below.

Figure 6 demonstrates the three-point bend test results for flexural strength of the studied mixes mortars at 7, 14, and 28 days. It is noted, as expected, that the incorporation of the rock waste was able to increase flexural strength for all evaluated cure times. This is probably due to the capacity of this residue to increase the density

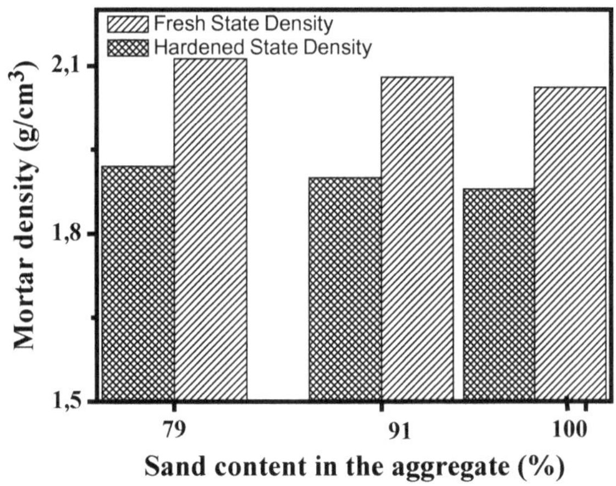

Fig. 5 Mortar density in fresh and hardened state of the mixes mortars studied

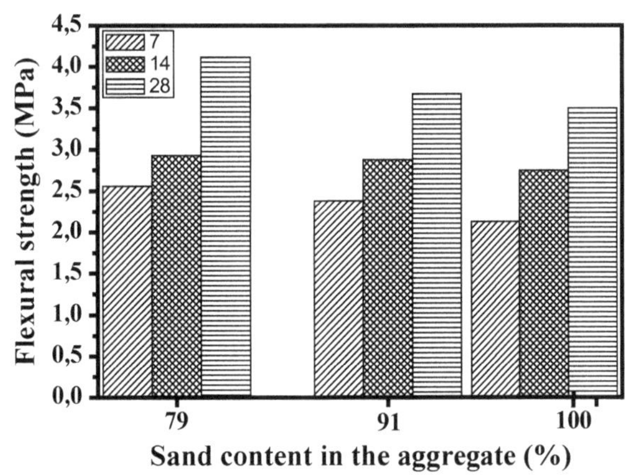

Fig. 6 Flexural strength at 7 days of the mixes mortars studied

in the hardened state of the mortar, reducing the voids, and increasing the contact points between the cementitious matrix and the aggregate. In addition, it is verified the progressive increase of the resistance with the increase of the curing time until the 28 days, evidencing the viability of the use of this residue as substitute for the sand.

Figure 7 shows that the increase of the percentage of rock waste caused an increase at the compressive strength, independently of the curing age. This behavior is ascribed to the reduction of the void ratio due to the improvement in the particle size distribution of the rock waste as partial substitute of sand as aggregate. The observed increase in both the flexural strength and compression strength with the incorporation of rock waste in the mortars was similar to that found by Rao et al. [19] when evaluating the performance of concrete with replacement of sand by granite residue.

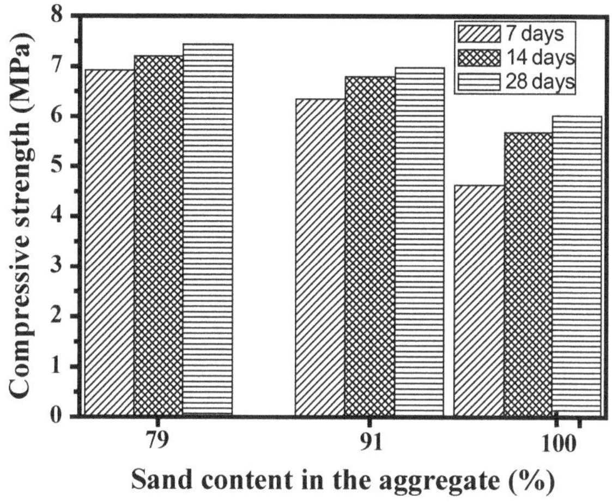

Fig. 7 Compressive strength at 7 days of the mixes mortars studied

Conclusions

The efficiency in the use of rock waste and Alfred's mathematical model for the calculation of grain size optimization is useful to improve mortar properties. The increase in both fresh body density and hardened state density through the application of this methodology improved bending and compressive stress resistance. In this way, an alternative methodology is proposed to better utilize solid wastes that have granular morphology in which the main objective is not only to provide an environmentally sustainable destination for the waste, but also to improve the properties of the mortar.

References

1. Ashraf WB, Noor MA (2011) Performance evaluation of concrete properties for different combined aggregate gradation approaches. Procedia Eng 14:2627–2634
2. Ergul Y, Yasin E, Alaettin K (2004) Effect of limestone aggregate type and water-cement ratio on concrete strength. Mater Lett 58:772–777
3. Ioannis PS, Konstantinos GT (2013) Effect of composition variations on bond properties of selfcompacting concrete specimens. Constr Build Mater 41:252–262
4. Mucteba U, Kemalettin Y, Metin I (2010) The effect of the initial water to cement ratio on shielding properties of ordinary concrete. Constr Build Mater 52:491–493
5. Rafat S, Paratibha A, Yogesh A (2012) Influence of water/powder ratio on strength properties of self-compacting concrete containing coal fly ash and bottom ash. Constr Build Mater 29:73–81
6. Ronnen L, Hashem A (2002) Effects of composition and exposure on the solar reflectance of Portland cement concrete. Cem Concr Res 32:1679–1698
7. Sari D, Pasamehmetoglu AG (2005) The effects of gradation and admixture on the pumice lightweight aggregate concrete. Cem Concr Res 35:936–942
8. Abdel-Jawad YJ, Abdullah WS (2002) Design of maximum density aggregate grading. Constr Build Mater 16:495–508

9. Paiva H, Velosa A, Ferreira VM (2015) Correlation between mortar and concrete behavior using rheological analysis. J Build Eng 4:177–188
10. Hüsken G, Brouwers HJH (2008) A new mix design concept for earth-moist concrete: a theoretical and experimental study. Cem Concr Res 28:1246–1259
11. Yu QL, Spiesz P, Brouwers HJH (2013) Development of cement-based lightweight composites—Part 1: mix design methodology and hardened properties. Cement Concr Compos 44:17–29
12. Yu R, Onna DV, Spiesz P, Yu QL, Brouwers HJH (2016) Development of ultra-lightweight fibre reinforced concrete applying expanded waste glass. J Clean Prod 112:690–701
13. Furnas CC (1929) Flow of gasses through beds of broken solids. Bureau of Mines Bulletin, p 307
14. Andreasen AHM, Andersen J (1930, in German) About the relationships between grain gradations and gap in products of loose grains (with some experiments). Kolloid-Zeitschrift 50:217–228
15. Funk JE, Dinger DR (1994) Predictive process control of crowded particulate suspensions, applied to ceramic manufacturing. Kluwer Academic Press, Boston
16. Moura WA, Gonçalves JP, Leite RS (2002) Use of marble and granite cutting residue in coating mortar and making floor tiles (in port.). Sittientibus 26:49–61
17. Vladimir GH, Graça V, Lourenço PB (2016) Workability and hardened properties of mortars. Constr Build Mater 25:2980–2987
18. MATTOS LRS (2001) Identification and characterization of external mortars used in the city of Belém-PA (in Port.). Master Dissertation, Universidade Federal do Rio Grande do Sul
19. Rao KB, Besai VB, Mohan DJ (2011) Experimental investigations on mode II fracture of concrete with crushed granite stone fine aggregate replacing sand. Mater Res 15:41–50

Technological Properties of Brick Waste-Based Geopolymer

Kátia Cristina P. Faria, Carlos Mauricio Fontes Vieira, Dylmar P. Dias, Marcos Yuri S. Fagundes and Weslley M. Ferreira

Abstract Studies using the geopolymerization process of the brick wastes from civil construction in the production of new materials, with characteristics and properties similar to traditional ceramics, has been growing in recent years. The present work used these brick wastes as precursor material and Na_2SiO_3 (alkaline sodium silicate) and KOH (potassium hydroxide) as alkaline activators on the geopolymer dosages, where the blend was determined with the molar ratio SiO_2/Al_2O_3 equal to 4.0. The molded and pressed specimens were left to mature at room temperature and at 60 °C during 7 and 14 days. Afterward, they were submitted to the flexural strength, linear shrinkage, water absorption, and apparent porosity tests. The results showed that, the longer the cure time, the better the technological characteristics of the specimens, both at room and controlled temperature. Thus, it was verified that the brick wastes are an interesting alternative as binder in the production of geopolymeric ceramics.

Keywords Alkaline activators · Brick wastes · Geopolymeric ceramics

Introduction

The alkaline activation of aluminosilicates is a practice used by man for several centuries, but is a technology that has not been widely disseminated throughout history and recently rediscovered through several studies. Some works, using artificially agglutinated stone materials such as the Coliseum of Rome, the Roman Pantheon, and the Baths of Caracalla, have exceptional durability, strength, and excellent mechanical properties [1]. The interest in alkaline or geopolymerized activated materials is part of the need to obtain new products with better mechanical and physical char-

K. C. P. Faria (✉) · C. M. Fontes Vieira · M. Y. S. Fagundes · W. M. Ferreira
State University of the Northern Rio de Janeiro – UENF, Advanced Materials Laboratory – LAMAV, Av. Alberto Lamego 2000, Campos dos Goytacazes, Rio de Janeiro 28013-602, Brazil
e-mail: katiacpfaria@hotmail.com

D. P. Dias
State University of the Northern Rio de Janeiro – UENF – Civil Engineering Laboratory - LECIV, Av. Alberto Lamego 2000, Campos dos Goytacazes, Rio de Janeiro 28013-602, Brazil

S. Ikhmayies et al. (eds.), *Green Materials Engineering*, The Minerals, Metals & Materials Series, https://doi.org/10.1007/978-3-030-10383-5_32

acteristics, more strength, durable, with lower production cost and mainly with less environmental impact in their production, being able to be used as alternative materials to concretes and mortars in civil construction [2].

Geopolymers are inorganic materials formed by chemical reactions between oxides of aluminosilicates and alkali metal solutions under alkaline conditions, with partial or total amorphous polymer structure containing Si–O–Al bonds [3]. The mechanical properties of the geopolymer structure are sensitive to particle size, chemical composition, SiO_2/Al_2O_3 ratio, temperature and curing time, and concentration of the activator solution. The type of alkaline activator and its concentration are very important parameters in the strength of the geopolymers. However, very high molarities may not be beneficial to determine these properties [4, 5].

The application of the geopolymer ranges from ceramic materials such as brick, tiles, mortars, cement, even highly complex materials such as aeronautics. A tendency to follow, in terms of energy and environment, is the method of ceramic material production through geopolymerization [6]. Although it is a practice not widely used yet, it is of great importance because it produces materials using low temperature, thus avoiding the emission of greenhouse gases [7].

The great advantage in the use of geopolymers is the possibility of incorporating waste, which are often been discarded or placed in deposits, producing permanent costs for the companies. Many residues can be used to produce geopolymers, such as fly ash, blast furnace slag, grounded glass, and grounded brick residue. [8]. The latter has proved its efficiency as a binder on the geopolymerization process. However, it is little used, and several studies investigating this residue as precursor in the synthesis of geopolymers is already a reality and has been highlighting in recent years [9–12].

Therefore, the objective of this study is to use the grounded brick residue as an activated binder by alkaline mixture of KOH and Na_2SiO_3, and to develop ceramics by the geopolymerization process. In addition, this work aims to evaluate the technological properties in flexural, water absorption, apparent porosity, and linear retraction, during 7 and 14 days cure at room temperature (R.T.) and at 60 °C.

Methodology

In this experimental investigation two geopolymer blends were prepared, one by pressing process and another one by molding. For the geopolymer mixtures, grounded brick residue was used as precursor material, coming from the ceramic industries of Campos dos Goytacazes-RJ. The residue presents chemical composition equal to 58% SiO_2 and 37% Al_2O_3. So, it was necessary to use Na_2SiO_3 acting as a complementary source of silica, and also as an activator for the geopolymerization process. The SiO_2/Al_2O_3 ratio of approximately 4 was the same for all dosages (Table 1), as patented by Davidovits [6]. The commercial aggregate material used was sand. The potassium hydroxide (85%) and powdered sodium silicate—Na_2SiO_3 (49% SiO_2 and 23% NaOH), were the alkaline materials used to fabricate the alkaline solution.

Table 1 Pressed (GP) and molded (GM) geopolymers, time and temperature of cure

Days	Room temperature (R.T.)		60 °C	
7	GM1	GP1	GM2	GP2
14	GM3	GP3	GM4	GP4

The alkaline solution was prepared using a molar concentration equal to 10 mol/L composed by flakes of potassium hydroxide, powdered alkaline sodium silicate, and distilled water. Considering the heat generated during the preparation, the alkaline solutions were cooled to room temperature during 24 h before its use.

The granulated geopolymer mass was prepared using solution/grounded residue brick + sand ratio of 0.26 and compacted by uniaxial pressure in a hydraulic equipment, using a rectangular steel mold (114.5 mm × 25.4 mm × 10.0 mm). The molded specimens were prepared using solution/grounded residue brick + sand ratio equal to 0.41 and poured into molds of 115.0 mm × 30.0 mm × 20.0 mm.

On the fabrication of the molded specimens, the entire volume of the alkaline solution was added to the sand and mixed manually during 5 min. The grounded residue brick was then added in an equal amount of the sand and mixed during another 5 min. The mixture was then poured into the molds. The same method was used for the granulated mixture produced by pressing. Because it was a semi-dry mass, it was necessary to pass it in a 10 mesh sieve to obtain a more homogenous mixture. The specimens were pressed using a pressure equal to 5.0 t. The molded and pressed samples were left to cure at room temperature for 24 h with their upper surfaces covered with a plastic film to prevent excessive loss of moisture. Immediately after this period, the molded specimens were removed from the molds and all samples remained at room temperature and at 60 °C until the tests (7 and 14 days).

After the curing period, the samples were submitted to some tests. The flexural strength was determined by a universal test machine, INSTRON model 5582, at a loading rate of 1.0 mm/min. In the tests of water absorption, the sample weights were measured after dry during 24 h and after immersion in water. The apparent porosity was made using a marte hydrostatic scale, model AD500. Linear shrinkage was calculated by measuring the difference in sample length before and after curing.

Results

Considering that there is no specific and available standard that deals with geopolymeric materials, similar to the object of study addressed in this work, an alternative was sought to supply this need. In this way, a bibliographic reference and technical standards of ceramic materials were used to evaluate the quality of the geopolymerized ceramics developed in this work.

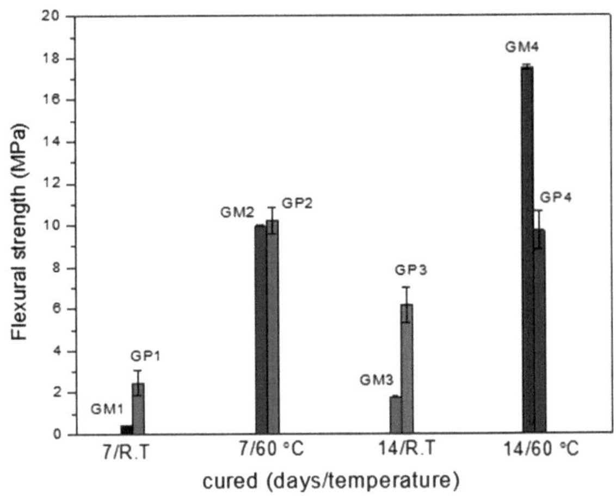

Fig. 1 Flexural strength of molded (GM1, GM2, GM3, and GM4) and pressed (GP1, GP2, GP3, and GP4) specimens at room temperature and 60 °C

Flexural Strength

Figure 1 shows the flexural strength of the geopolymerized specimens. There was an increase on strength with curing time for most of the specimens. It was observed that all samples cured at 60 °C had the highest values. Based on Santos [13], the minimum bending strength required for the manufacture of masonry bricks is 2.0 MPa, 5.5 MPa for perforated bricks and 6.5 MPa for tiles. In general, most of the samples reached the values of this reference, and only GM1 and GM3 did not have the strength expected, probably because the time and cure temperature were insufficient for complete geopolymerization of the samples. The highest strength value was observed for GM4, because the temperature and curing time were sufficient to accelerate the reactivity. As for the specimens pressed, the flexural strength was greater in the majority, showing that the degree of compaction influenced this increase. However, the prolonged curing time at controlled temperature did not influence the flexural strength gain, as seen in GP4 in comparison with GP2.

Linear Shrinkage

Figure 2 shows the linear shrinkage of all samples. As expected, the shrinkage increases with an increase of the time and temperature of cure. The molded specimens presented values of shrinkage higher than the pressed ones. This is due to the water content that for these samples were higher. In the pressed specimens, in addition to the lower water content, the degree of compaction may have influenced the low linear retraction. The shrinkage values of all pressed samples follow the dimensions determined by the specific standard for tiles [14], where the allowed dimensional tolerance is ±2.0%. It was observed that GP2 and GP4 present very close linear

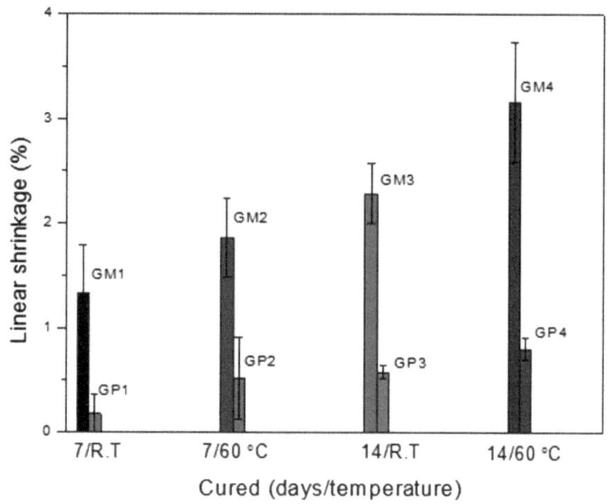

Fig. 2 Linear shrinkage of the molded (GM1, GM2, GM3, and GM4) and pressed (GP1, GP2, GP3, and GP4) specimens at room temperature and 60 °C

shrinkage values, taking into account the standard deviation. These results corroborate the flexural strength values which were constant for these samples. This fact is related to the water content in the GP4 sample that was insufficient to complete the geopolymerization and to obtain an expected flexural strength in relation to its cure time.

Water Absorption and Apparent Porosity

The water absorption and the apparent porosity of the specimens are shown in Figs. 3 and 4. It was observed that the absorption is related to the open porosity of the material. For the samples GM1 and GP2, it was not possible to determine these properties because during the analysis the samples deteriorate, proving that the geopolymerization process was low and did not produce a strength matrix. Note that most of the samples obtained low values of absorption and porosity, and only the GM3 and GP3 samples presented higher values, indicating a structure of low strength and high porosity. The GP2 and GP4 samples obtained slightly higher values compared to GM2 and GM4, indicating that the degree of compaction influences the formation of the open pores. It was observed that GM4 obtained the lowest values of absorption and porosity of all specimens, corroborating the high flexural strength and low shrinkage. Most of the presented results of water absorption do not fit the standard limits for bricks [15] that should be between 8.0 and 22.0%. Only GM3 and GP3 reflected this behavior. However, all other specimens, pressed and molded, fit on the tile standard [14], which should not exceed 22.0%.

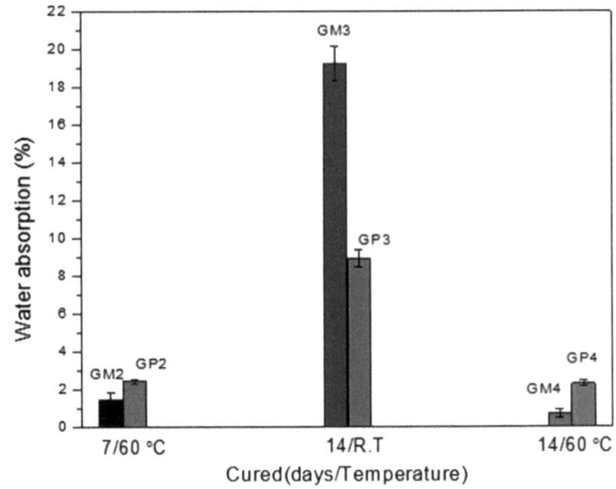

Fig. 3 Water absorption of molded (GM2, GM3, and GM4) and pressed (GP2, GP3, and GP4) at room temperature and 60 °C

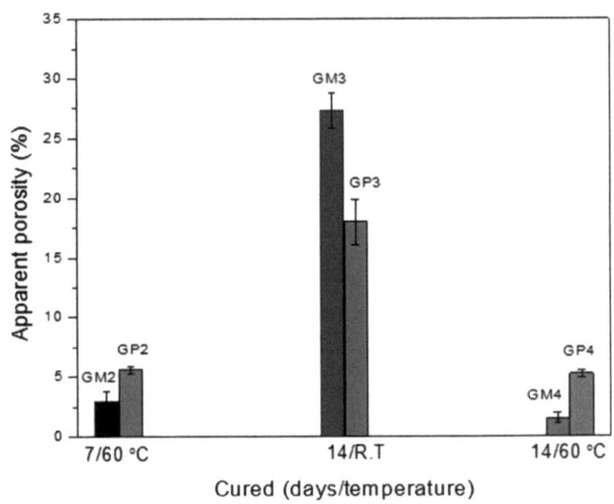

Fig. 4 Apparent porosity of molded (GM2, GM3, and GM4) and pressed (GP2, GP3, and GP4) at room temperature and 60 °C

Conclusions

This study showed that ceramic materials obtained by the geopolymerization process of brick residue are technically feasible. The technological properties of the developed geopolymer are compatible with the ceramics obtained by the conventional burning process. It was verified through the experimental results that the compaction, the water content, and the controlled temperature are determinant to improve the characteristics of the geopolymerized roofing tiles and bricks. In the case of specimens subjected to the curing process at room temperature, a longer curing time should be used to produce geopolymeric roofing tiles and bricks with better technological characteristics. Therefore, considering the conditions and methods used, as well as the normative references, it was concluded that the pieces produced in this work can be used in the civil construction area.

References

1. Pinto AT (2006) Introdução ao Estudo dos Geopolímeros. Universidade de Trás-dos-Montes e Alto Douro, Vila Real
2. Yun-Ming L, Cheng-Yong H, Li L, Jaya NA, Abdullah MMAl, Jin TS, Hussin K (2017). Formation of one-part-mixing geopolymers and geopolymer ceramics from geopolymer powder. Constr Build Mater 156:9–18
3. Davidovits J (1999). Chemistry of geopolymeric systems, terminology. In: Proceedings of the 2nd international conference on geopolymer'99. Saint-Quentin, France, 30 de June, 1–2 de July de 1999
4. Cioffi R, Maffucci L, Santoro L (2003) Optimization of geopolymer synthesis by calcination and polycondensation of a kaolinitic residue. Resour Conserv Recy 40:27–38
5. Hardjito D, Rangan BV (2005). Development and properties of low-calcium fly ash based geopolymer concrete. Research report GC, 1, Australia, Perth, Faculty of Engineering, Curtin University of Technology, pp 1–94
6. Davidovits J (2008) Geopolymer chemistry and applications, 2nd edn. France
7. Almeida KS, Roberto ALS, Camila SM (2014). Análise dos Impactos Ambientais Gerados pela Indústria de Cerâmica Vermelha no Piauí. Cerâmica Industrial. pp 33–34
8. Araújo L, Meert RLB, Labrincha JA, Senff L (2017). Desenvolvimentos de geopolímero a partir de resíduo da construção como agregado. 8° forum internacional de Resíduos Sólidos, UTFPR, Coritiba, Brasil
9. Reig L, Tashima MM, Borrachero MV, Monzó J, Cheeseman CR, Payá J (2013). Properties and microstructure of alkali-activated red clay brick waste. Constr Build Mater, 98–106. Elsevier
10. Komnitsas K, Zaharaki D, Vlachou A, Bartzas G, Galetakis M (2015). Effect of synthesis parameters on the quality of construction and demolition wastes (CDW) geopolymers. Adv Powder Technol, 368–375. Elsevier
11. Zawraha MF, Gadoa RA, Feltin N, Ducourtieux S, Devoille L (2016). Recycling and utilization assessment of waste fired clay bricks (Grog) with granulated blast-furnace slag for geopolymer production. Process Saf Environ Prot, 237–251
12. Azevedo AG de S, Strecker K, Lombardi CT (2018). Produção de geopolímeros à base de metacaulim e cerâmica vermelha. Cerâmica 64:388–396
13. Santos PS (1989) Ciência e Tecnologia das Argilas, 2nd edn. Edgard Blücher, São Paulo
14. ABNT - Associação brasileira de normas técnicas - NBR 15270-1 (2005) Componentes cerâmicos. Parte 1: Blocos cerâmicos para alvenaria de vedação – Terminologia e requisitos
15. ABNT - Associação brasileira de normas técnicas - NBR 15310 - (2009). Componentes Cerâmicos – Telhas – Terminologia, requisitos e métodos de ensaio

Use of Waste of Ornamental Stone in Ceramic Mass Incorporation in Brazil

Maria Angélica Kramer Sant'Ana, Mônica Castoldi Borlini Gadioli, Michelle Pereira Babisk and Carlos Mauricio Fontes Vieira

Abstract Due to the great generation of waste in the industries of ornamental stone that can cause damage to the environment, it was necessary to carry out studies for their use. Therefore, one of its uses that has been studied is the incorporation of these wastes in ceramic masses. Aiming to mitigate the impacts caused by the disposal of these wastes and minimize the use of the raw material (clay). The objective of this work was to study the use of sawdust waste of blocks of ornamental stone in industrial ceramic mass. Characterization of the waste and ceramic mass was carried out. Residual ceramic (0%) and 10, 20, 30, 40, and 50% by weight of the waste from the sawdust were prepared. Test specimens were formed by uniaxial pressing, dried at 110 °C and sintered at 950 °C. Technological tests were carried out on the specimens for determination of apparent specific mass, water absorption, dilatometry, and linear retraction. With all these tests carried out on the materials and test specimens, it was possible to observe that the use of the wastes in the incorporation of ceramic masses is possible and can even improve the ceramic artifact, bringing a sustainable development and industrial ecology to both sectors involved.

Keywords Stone waste · Use waste · Ornamental stone

M. A. K. Sant'Ana (✉) · M. C. B. Gadioli
CETEM/MCTIC, Mineral Technology Center, Regional Nucleus of the Espirito Santo, Rodovia Cachoeiro – Alegre, Km 05, Cachoeiro de Itapemirim, ES 29311-970, Brazil
e-mail: mariaangelicaks@gmail.com

M. C. B. Gadioli
e-mail: mborlini@cetem.gov.br

M. P. Babisk · C. M. F. Vieira
Laboratory for Advanced Materials, State University of Northern Rio de Janeiro, Campos dos Goytacazes, RJ, Brazil
e-mail: michellebabisk@hotmail.com

C. M. F. Vieira
e-mail: vieira@uenf.br

S. Ikhmayies et al. (eds.), *Green Materials Engineering*, The Minerals, Metals & Materials Series, https://doi.org/10.1007/978-3-030-10383-5_33

Introduction

The ceramic masses are often composed of a mixture of clays, and the clays in turn are minerals found abundantly in the nature. It is a material used from the beginnings of humanity to the present day for the manufacture of diverse ceramic products, like bricks, floor, and tiles among several other industrial applications [1]. The demographic growth and development of the South and Southeast regions of the country (Brazil) have made the ceramic sector also to increase [2]. There are approximately 6,903 Ceramics and Olarias companies in Brazil, generating around 18 million annual revenues [3]. In controversy, many of the ceramic industries in Brazil still use outdated techniques for the manufacture of ceramic artifacts, within other obstacles that hinder the improvement of the ceramic industrial sectors [4].

However, the ornamental stone sector is of great importance for Brazil and is in 4th place in the world ranking of higher production [5]. The ornamental stones are removed in blocks and usually pass through a processing stage, where occurs the sawing by multi-wire or multi-blade of the blocks for the production of slabs and the subsequent polishing of these slabs. About 26% of a block becomes waste during the sawing stage [6]. The extraction and processing of ornamental stone is an activity that generates too much solid waste. It is estimated that in Brazil more than 22 Mt of waste was generated in 2012, of which approximately 20 Mt is of residues generated in the quarries themselves and approximately 2 million tons of fine residues per year [7].

Due to too much amount of stone waste generated and discarded annually, it was necessary to carry out researches seeking the technical-economical feasibility for its use as raw material by industries, mainly in the civil construction sector, such as the ceramics industry (case studied in this article), industry concrete, among other uses [7, 8].

Studies like these seek to study technological and economic viability for both sectors involved, aiming to promote sustainable development, thereby promoting the well-being of present and future generations.

In order to contribute to the mitigation of the environmental impact and consequently, with the sustainable development, the objective of this work was to study the use of sawdust waste of blocks of ornamental stone in industrial ceramic mass.

Materials and Methods

The materials used in this study were the raw materials: waste of sawdust collected at a company in Cachoeiro de Itapemirim-ES and industrial ceramics mass collected at a company in the region of Paineiras, Itapemirim-ES.

Samplings were performed according to the solid waste sampling standard NBR 10007 [9]. The waste and ceramic mass were dried, sieved, sampled again, and then characterized.

X-ray diffraction (XRD), chemical analysis, and particle size distribution analyses were performed in the mass and in the waste.

The chemical composition was determined by X-ray fluorescence (XRF) in a Spectrometer WDS model AXIOS from Panalytical and the loss by calcination in a Leco TGA-701 equipment. The mineralogical analysis was conducted by X-ray diffraction (XRD) in a D4 model Endeavor Brucker diffractometer coupled with a Goebel mirror for parallel beam. The XRD operation was carried out with CoK radiation with the 2θ angle ranging from 5 to 80°. The particle size distribution was determined using the Mastersizer 2000 equipment, from Malvern Instruments. Micrograph of the waste was obtained by SEM using the FEI Quanta 400 microscope.

In order to verify the mobility of pollutants, leaching and solubilization tests, according to NBR 10005 [10] and 10006 [11], were performed in the waste. The elements, as well as other substances required by standard NBR 10004 [12], were determined in the solubilization and leaching extracts.

The following compositions were prepared in this study: ceramic mass without addition of waste, denominated 0% and incorporations of 10, 20, 30, 40, and 50% by waste in the ceramic mass, totaling 6 different masses.

After being prepared, these masses were homogenized in a porcelain jar mill and moistened to 8% water, sifted and then shaped to form test bodies by 10-tonne uniaxial pressing using a matrix of steel with dimensions of 114 × 25 mm and approximately 60 g of material. In total, 30 test bodies were prepared.

After being ready, the specimens obtained were oven dried at 110 °C. And, after drying, the apparent specific mass was calculated according to standard C373-72 [13]. The specimens were sintered at temperature of 950 °C with a heating rate of 2 °C/min.

Technological tests were performed to determine specific mass, linear shrinkage, dilatometry, and water absorption. The linear shrinkage was obtained by measurements of the length of the specimens before and after sintering using a caliper. The water absorption test was done according to the norm NBR 15270-2 [14].

Results and Discussion

The chemical composition of the raw materials is shown in Table 1, giving evidence to the fluxing oxides ($K_2O + Na_2O$) that can allow an improvement in the densification of the ceramics and assist in their sintering. The fluxing oxides are present and in a relevant amount in the composition of the sawdust waste, while in the ceramic mass it presents low content.

The minerals found in the XRD test were quartz, microcline, albite, and muscovite for the sawdust waste, and found quartz, kaolinite, vermiculite, muscovite, gibbsite, and sepiolite for the ceramic mass used.

By analyzing the particle size distribution with the results shown in Table 2, the ceramic mass and the waste present very close classification with 10% of the particles

Table 1 Chemical composition of raw materials

Materials	Na_2O	MgO	Al_2O_3	SiO_2	P_2O_5	K_2O	CaO	TiO_2	Fe_2O_3	[a]LoI
Waste	5.50	0.46	17.00	69.90	0.14	3.40	1.50	0.24	1.30	0.55
Ceramic mass	0.17	0.93	31.20	47.40	0.21	0.80	0.26	1.10	6.00	11.60

[a]Loss on ignition

Table 2 Particle size distribution of the raw materials

Materials	d(0.1) (μm)	d(0.5) (μm)	d(0.9) (μm)
Ceramic mass	2.378	14.341	57.298
Waste	2.241	13.040	70.167

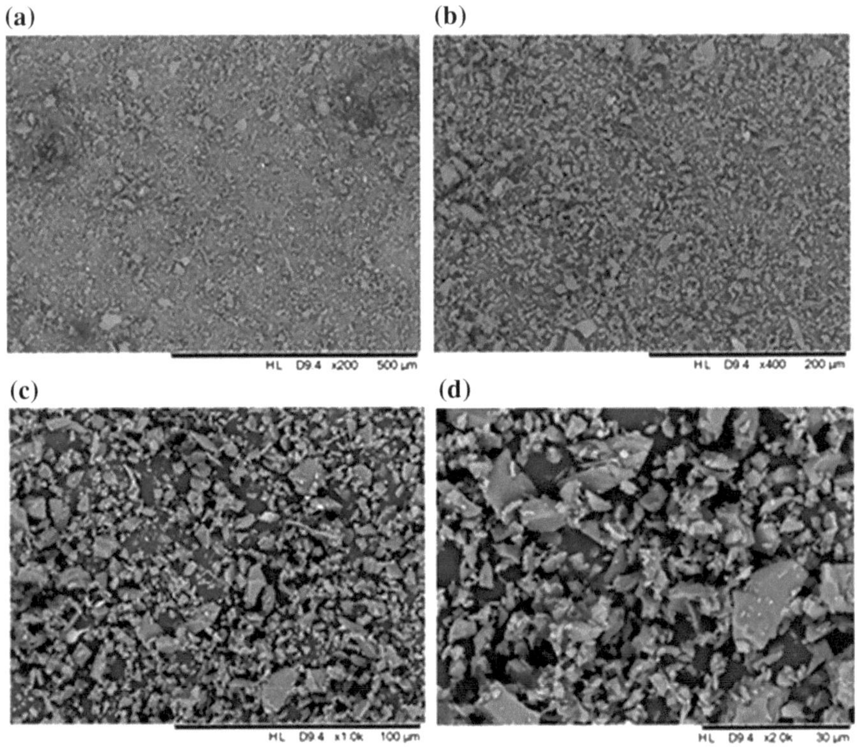

Fig. 1 Micrographs of the waste. **a** 200x, **b** 400x, **c** 1000x, **d** 2000x

in the clay phase. It is observed that 50% of the particles are below 15 μm for both raw materials and that 90% of the particles are below 71 μm.

The SEM micrographs shown in Fig. 1 have particles of varying sizes, as already noted in the particle size distribution assay. And, it is also noticed that the grains are irregular and angular, and this is a characteristic of the stone waste in general.

Table 3 Analysis of the leaching extract parameters

Parameters	Maximum limit in extract (mg/L) NBR 10004 standard [12]	Leaching extract of the waste (mg/L)	Results
Arsenic	1	0.02	Inside the limit
Barium	70	0.73	Inside the limit
Cadmium	0.5	<0.0004	Inside the limit
Lead	1	0.03	Inside the limit
Total chrome	5	<0.001	Inside the limit
Mercury	0.1	<0.009	Inside the limit
Silver	5.0	<0.005	Inside the limit
Selenium	1	<0.008	Inside the limit

Table 4 Analysis of the solubilization extract parameters

Parameters	Maximum limit in extract (mg/L) NBR 10004 standard [12]	Solubilization extract of the waste (mg/L)		Results
Aluminum	0.2	3.7	4	Out of limit
Arsenic	0.01	<0.008	<0.008	Inside the limit
Barium	0.7	0.27	0.25	Inside the limit
Cadmium	0.005	<0.0004	<0.0004	Inside the limit
Lead	0.01	<0.008	<0.008	Inside the limit
Copper	2	<0.01	0.01	Inside the limit
Total chromium	0.05	<0.001	<0.001	Inside the limit
Iron	0.3	0.16	0.22	Inside the limit
Manganese	0.1	0.09	0.03	Inside the limit
Mercury	0.001	<0.009	<0.009	Inside the limit
Silver	0.05	<0.005	<0.005	Inside the limit
Selenium	0.01	<0.008	<0.008	Inside the limit
Sodium	200	18.8	19.2	Inside the limit
Zinc	5	0.06	0.07	Inside the limit

The results of solubilization and leaching of the waste are shown in Tables 3 and 4. For the results of the leaching analysis, all parameters are within the limits required by the standard [12].

In the solubilization tests, carried out in duplicates, in relation to the analyzed elements, only the parameter of aluminum is above the maximum limit allowed by the standard, classifying the waste as non-inert waste class II A [12].

Figure 2 shows the apparent specific mass of the compositions prepared without and with sawdust waste. The technological characteristic of apparent specific mass

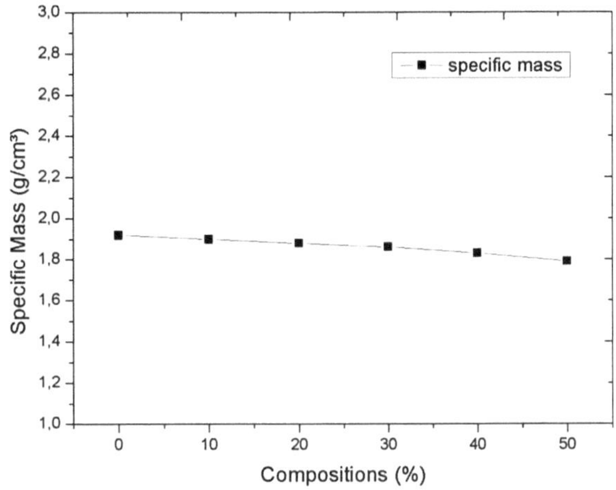

Fig. 2 Specific mass of compositions (g/cm^3)

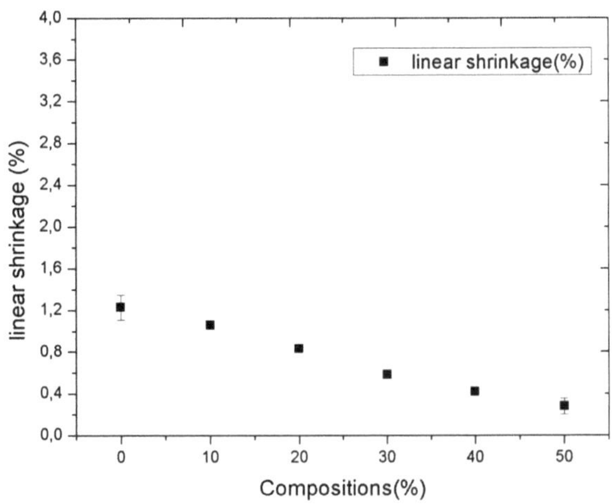

Fig. 3 Linear shrinkage of compositions (%)

is very important, this is because it indicates the degree of packaging of the particles. There was a slight decrease in the value of the specific mass as the percentage of waste increased, indicating that the waste influenced the degree of packaging of the ceramic mass, so the burning behavior of these compositions can be influenced by the degree of porosity different from dry species.

It can be seen in Fig. 3 that linear shrinkage decreases with increasing waste percentage. The low linear shrinkage contributes to a better dimensional control of the material.

The water absorption values presented in Fig. 4 remained the same or close to those of the waste-free mass up to values of 30% of incorporation of waste. The lower the water absorption, the lower the open porosity of the specimens.

The results of dilatometry (Figs. 5 and 6) for the compositions of 0 and 50% of waste incorporation were done with the ends to better observe the stages of dilations

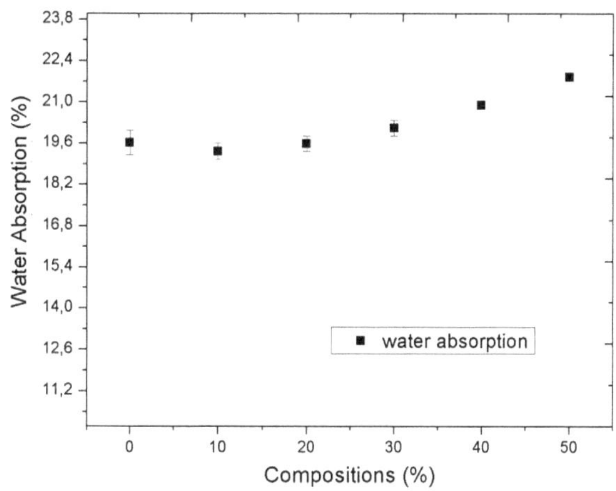

Fig. 4 Water absorption of compositions (%)

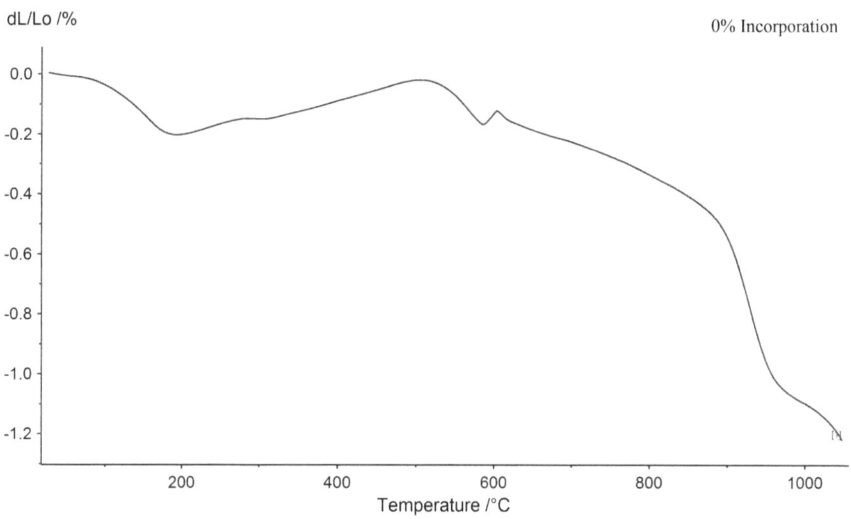

Fig. 5 Dilatometry of the pure ceramic mass (%)

of the submitted temperatures. It can be noticed that for both results there is a water loss retraction in the 100–200 °C range, after which the graphs differ completely. In the composition 0% there was a greater retraction, mainly, after the range of 600 °C. At 50%, there was an increasing expansion to 900 °C (approximately). It is observed that for the burning temperature studied in this work, the variation of dilation is decreasing for the compositions.

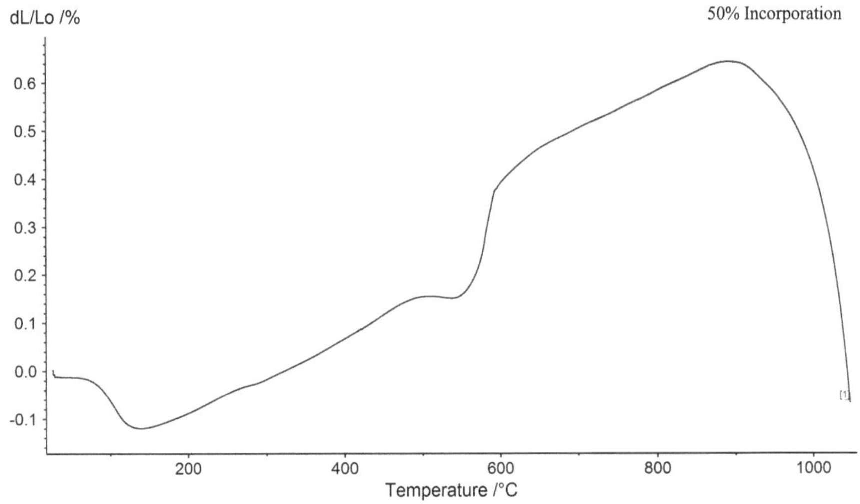

Fig. 6 Dilatometry of the composition with 50% incorporation of waste (%)

Conclusions

- It was possible to observe that, in general the results obtained were promising.
- The waste has in their composition the presence of alkaline oxides known as the fluxing agents that contribute to the ceramic sintering. In the XRD of the waste, feldspar was found.
- The ceramic mass and the studied waste have a very similar particle size distribution.
- The waste is classified as not inert. But, during the burning of the ceramic, the constituents can be fixed in the ceramic matrix.
- Results promising were found for the water absorption. For the compositions with incorporation of up to 30% of waste practically not have influence of the waste in the results when compared to the ceramic mass without waste, taking into account the bar of statistical error.
- Therefore, it is observed that the use of the wastes in the incorporation of ceramic masses is possible and contributes to the reduction of wastes to be discarded in the nature, mitigation of extracted raw material (clay), improvement of the properties of the ceramics and the sustainable development and industrial ecology of these sectors.

Acknowledgements The authors thank the National Council for Scientific and Technological Development, CNPq, and Foundation for protection of Research and Innovation in Espírito Santo, FAPES, for the support of the research project.

References

1. Manoharan C et al (2012) Ceramic 58:412–418
2. ABCERAM—Brazilian association of ceramics (2018, in Portuguese) Ceramics in Brazil-general considerations. https://abceram.org.br/consideracoes-gerais/. Accessed May 2018
3. ANICER—national association of ceramics industry (2018, in Portuguese). ANICER - sector data. https://www.anicer.com.br/anicer/setor/. Accessed September 2018
4. Amaral LF et al (2018) Development of ceramic paver with ornamental rock waste. J Mater Res Technol. https://doi.org/10.1016/j.jmrt.2018.05.009
5. ABIROCHAS—Brazilian Association of the Ornamental Stone Industry (2018, in Portuguese) Balance of Brazilian exports and imports of ornamental stones in 2017. Asa Sul-Brasília/DF. Report 01/2018. http://www.abirochas.com.br. Accessed September 2018
6. Silveira LLL et al (2013, in Portuguese) Improvement of ornamental stones. In: Francisco WHV et al (eds) Ornamental stone technology: research, mining and processing. Mineral Technology Center, Rio de Janeiro, RJ, Brazil, p 358. cap 7
7. Campos AR et al (2013, in Portuguese) Waste: treatment and industrial applications. In: Francisco WHV et al (eds) Ornamental stone technology: research, mining and processing. Center for Mineral Technology, Rio de Janeiro, RJ, Brazil, pp 435–436. cap 9
8. Digvijay SC et al (2017) Utilization of granite slurry waste in concrete: a review. Indian J Sci Technol 10(6):1–9. http://www.indjst.org. Accessed September 2018
9. ABNT—Brazilian norm association (2004, in Portuguese) NBR 10007. Solid waste sampling
10. ABNT—Brazilian norm association (2004, in Portuguese) NBR 10005. Procedure for obtaining leach extract from solid wastes
11. ABNT—Brazilian norm association (2004, in Portuguese) NBR 10006. Procedure for obtaining solubilized solid waste extract
12. ABNT—Brazilian norm association (2004, in Portuguese). NBR 10004. Solid waste - classification
13. ASTM—American society for testing and materials (1972) Water Absorptin, Bulk Density, Apparent Porosity and Apparent Specific Gravity of Fired Whiteware Products, C373-72, USA
14. ABNT—Brazilian Norm Association (2017, in Portuguese). NBR 15270-2. Ceramic blocks for structural masonry

Author Index

S. Ikhmayies et al. (eds.), *Green Materials Engineering*, The Minerals, Metals & Materials Series, https://doi.org/10.1007/978-3-030-10383-5

Subject Index

S. Ikhmayies et al. (eds.), *Green Materials Engineering*, The Minerals, Metals & Materials Series, https://doi.org/10.1007/978-3-030-10383-5

MIX
Papier aus verantwortungsvollen Quellen
Paper from responsible sources
FSC® C105338

If you have any concerns about our products,
you can contact us on
ProductSafety@springernature.com

In case Publisher is established outside the EU,
the EU authorized representative is:
Springer Nature Customer Service Center GmbH
Europaplatz 3, 69115 Heidelberg, Germany

Printed by Libri Plureos GmbH
in Hamburg, Germany